P9-BIN-798

MAMMALS
OF THE WORLD

There are many ways in which a group of animals may achieve success. The mammals are successful primarily because they are warm-blooded. As a result they are to some limited extent independent of the climate in which they live and are more active than large, cold-blooded animals can be. Unlike the other warm-blooded vertebrates, the birds, mammals are adapted for many widely varying forms of activity. As a result they are a very diverse group of animals in almost every possible way.

In this book an attempt has been made to bring out these points. The first part, which describes some of the essential features of mammals, lays particular emphasis on their warm-bloodedness. The remainder of the book describes them family by family, except in the cases of the bats and the rodents which are too numerous to permit this treatment. By focusing as much attention on small and unusual families as on large and successful ones this approach brings out the mammals' incredible diversity.

In this book, the region in which a mammal normally lives is included in the captions for many of the mammals pictured. The zoogeographical regions explained on Page 48 are used for mammals with a wide distribution; for those with a more limited range, a specific region is given.

A
GROSSET
ALL-COLOR GUIDE

MAMMALS OF THE WORLD

BY MICHAEL BOORER
Illustrated by John Beswick

GROSSET & DUNLAP
A NATIONAL GENERAL COMPANY
Publishers • New York

THE GROSSET ALL-COLOR GUIDE SERIES
SUPERVISING EDITOR ... GEORG ZAPPLER

Board of Consultants

RICHARD G. VAN GELDER · Chairman and Curator, Mammals, American Museum of Natural History

WILLIAM C. STEERE . Director, New York Botanical Gardens

SUNE ENGELBREKTSON · Space Science Co-ordinator, Port Chester Public Schools

JOHN F. MIDDLETON · Chairman, Anthropology, New York University

CARL M. KORTEPETER · Associate Professor, History, New York University

MICHAEL COHN · Curator, Cultural History, Brooklyn Children's Museum

FRANK X. CRITCHLOW · Consulting Engineer, Applied and Theoretical Electronics

Copyright © 1971 by Grosset & Dunlap, Inc.
All Rights Reserved
Published Simultaneously in Canada
Copyright © 1970 by The Hamlyn Publishing Group Ltd.
Library of Congress Catalog Card No.: 70-134998
ISBN: 0-448-00860-2 (Trade Edition)
ISBN: 0-448-04158-8 (Library Edition)
Printed in the United States of America

CONTENTS

THE EVOLUTION OF MAMMALS

Animal life first appeared on this planet about two billion years ago. The first animals were small invertebrates with bodies simple in structure. Some animals have remained comparatively simple and others have become more complicated, but most invertebrates have remained small. Many animals are microscopic, and to see some clearly, a magnifying glass is necessary. Animals have come to vary a great deal among themselves, and they consist of a whole range of major groups or phyla.

The first animals with backbones (vertebrates) evolved from invertebrate ancestors perhaps 500 million years ago. Gradually they became active animals of larger than average size—even minnows are larger than most invertebrates. The development of a backbone gave support to the bodies of some of these animals and made an active life on land possible for their descendants.

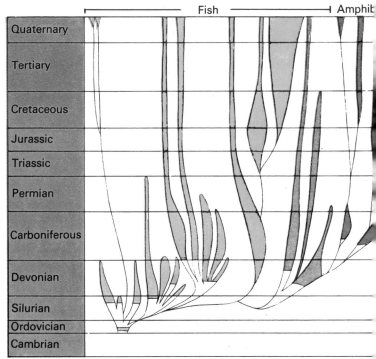

Vertebrate Evolutionary Relationships

Vertebrate Evolutionary Relationships—the width of the colored area in each group is proportional to the number of species. Geologic ages are shown on both sides of the chart. The numbers

The first vertebrates were fish and were adapted for life in water. From some of the early fish, the amphibians evolved; paired fins became limbs, swim-bladders became lungs. Although they still bred in water the amphibians could now begin to colonize the land. From them, the first reptiles evolved about 250 million years ago. These reptiles could even breed out of water, and although cold-blooded (like their fish and amphibian ancestors), they thrived on land. Birds are descended from the same ancestors as some of the reptilian dinosaurs.

The mammals are another separate branch from an even earlier reptilian stock. They evolved relatively slowly but by about 70 million years ago, as the great reptiles declined, mammals were becoming increasingly important. They had warm blood, and because of this they were well suited to living active lives in a wide variety of climates.

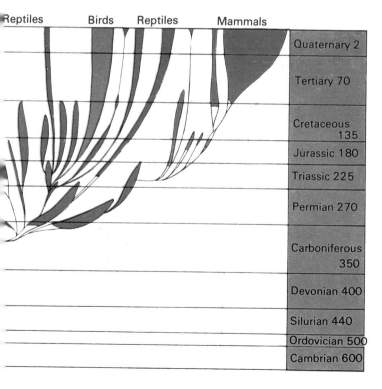

on the far right indicate (in millions of years) how far back these ages occurred.

The fossil evidence

The most direct evidence we have about the evolution of the mammals is provided by fossils. Most dead animals decay and eventually disappear, but occasionally the bones are preserved by exceptional conditions, such as by being engulfed in a swamp or in the sediments of shallow seas. In these and other ways fossils form, and enough have been found to give us a somewhat patchy picture of the past.

The early reptiles looked like short-limbed lizards. From these descended the great dinosaurs of the past, as well as the tortoises, crocodiles, lizards and snakes which still survive. Another line of descent led to a group known as the mammal-like reptiles.

The first mammal-like reptiles were still roughly lizard-shaped, and they were mostly meat-eaters. Gradually they began to resemble modern mammals, and by about 175 million years ago they were among the most important of land animals. By this time they had teeth not unlike those of modern mammals and may have had warm blood, hair and other purely mammalian features. Then they began to disappear.

Some small mammal-like reptiles lingered on, and by 100 million years ago they had given rise to some animals which were undoubtedly mammals. These were insect-eaters and in appearance were not unlike the shrews that we know today, except that they were larger. After this time the earth

Mammal-like reptile (*foreground*)

6

Early herbivorous mammals—uintatheres (*background*) and *Hyracotherium*, formerly eohippus (*foreground*).

went through a period of upheaval. There was much volcanic activity, and the climate underwent a considerable change. These events must have affected the world's animal population, and by about 70 million years ago the great reptiles had vanished and the mammals had become the most important land vertebrates. Already the ancestors of many of the main groups of mammals existing today had become recognizable.

Phenacodus, an early ungulate, still shows carnivore-like features, such as the long body and tail, but has tiny hoofs and broad cheek teeth.

Pygmy shrew and workhorse's
hoof to same scale

Blue whale and workhorse to
same scale

The incredible variety of mammals

Mammals can be defined as warm-blooded vertebrates with
hair which feed their young on milk. They are successful and
numerous animals because their basic design is well suited
to a whole range of ways of surviving. Almost any mode
of life in which great activity is important is possible to
them and, in the course of evolution, various groups of
mammals have become adapted in ways which exploit this
situation. In this process the mammals have become an
incredibly varied class of animals. They are far more varied
than some other groups of comparable size, such as the
birds, a group which has specialized in only one form of
activity—flight.

Some mammals scurry through thick undergrowth, while
others run in the open. Others burrow, climb trees in a
variety of ways, engage in prolonged leaps from one tree to
another, or fly with a skill which equals that of birds. Most
mammals can swim with varying degrees of proficiency;
some are superb swimmers, spending their entire lives in

water. Mammals can inhabit tropical deserts or frozen tundras. They can feed on plant material, meat or a mixture of the two. Some have become specially adapted to feed on insects; others are adapted to drink blood. Some mammals have special armor as a means of self-defense. Others are small and shy, and they avoid trouble by staying out of sight as much as possible. Others rely on speed to out-distance any pursuer, while some have become so large that their size is a defense in itself.

The range in size shown by members of the group is staggering. The smallest of mammals—some of the shrews—weigh only a fraction of an ounce, while the largest—some of the whales—weigh roughly 50 million times as much. The differences in shape between a stealthy hunter, such as a tiger, and a herbivore which runs on open plains and browses on the bark and leaves of trees, as does the giraffe, or the differences between the burrowing mole and the aquatic sea lion are equally startling. Yet each of these animals is a mammal, and they all share the essential features of the group.

Tiger

Giraffe

Mole

Sea lion

9

Skeleton of tree shrew

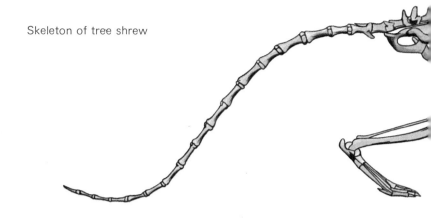

THE STRUCTURE OF MAMMALS

Because mammals are quite often large and because they usually live out of water they need a strong skeletal framework.

The head
The skull includes a bony box, the cranium or brain case, which protects the delicate brain. Embedded in the sides of the cranium are bony chambers which contain part of the ear apparatus; the front forms the support for the facial region which bears the nose and eyes. Beneath the facial region and also joined to the cranium is the tooth-bearing upper jaw. The lower jaw also bears teeth and is hinged. Unlike that of the reptiles the lower jaw of the mammals consists of only one bone on each side, the dentary, and this is undoubtedly a stronger arrangement.

The backbone
The vertebral column or backbone provides support for the body as a whole and protects the main nerve, the spinal cord, which passes through openings in each of the individual vertebrae. The vertebral column can be divided into a number of regions. Immediately behind the head are the cervical or neck vertebrae. In almost all species of mammals there are seven of these. The only exceptions are some of the sloths and the manatees.

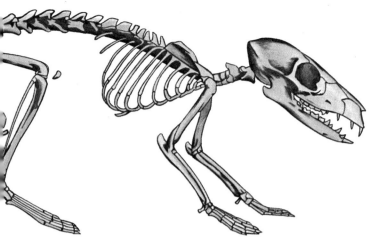

The thoracic vertebrae follow. In the mammals these are the only vertebrae to bear ribs. Immediately behind the rib-cage of a mammal is the lumbar region, the vertebrae of which are relatively large and bear prominent bony processes for the attachment of muscles. In most vertebrates the back-bone is very flexible, but in mammals flexibility is limited, the lumbar region being the only very flexible part. Even here the movement which takes place is not so much from side to side, as it is in the fish, amphibians and reptiles, but up and down.

Behind the lumbar region is the sacral region in which the vertebrae are fused together to form the rigid sacrum. To this is joined the pelvis which is made up of the hip bones.

The tail
A mammal's tail does not taper gradually like that of a reptile but is usually markedly slender for the whole of its length. The numbers of vertebrae in all parts of the vertebral column vary to some extent—even in the cervical region there are exceptions to the normal mammalian number of seven vertebrae, as has been mentioned—but the numbers of caudal or tail vertebrae vary more widely than any others. Some mammals have lost their tails in the course of evolution. Man, for example, has only three or four caudal verte-brae left, while some of the false vampire bats of the Old World have none at all. At the other end of the scale the long-tailed tree pangolin has 49 caudal vertebrae.

The muscles, skin and organs

The bones of a living mammal are moved by muscles. Every movable joint requires at least two muscles, one to pull it in a certain direction and another to pull it back again. Body movements are usually more complicated than this, and more than one pair of muscles may therefore be involved. As the number of joints in a mammal's body is large it is not surprising that the muscles are numerous and form a very significant part of the total body weight.

The skin consists of two layers. The inner layer, or *dermis,* holds the body together and provides a defense against harmful bacteria. It contains many blood vessels and sensitive nerve endings as well as the various skin glands. The outer layer, or *epidermis,* provides mechanical protection and consists, at the surface, of a dead layer of protein material known as keratin. Various mammalian structures which

Internal organs of a mammal

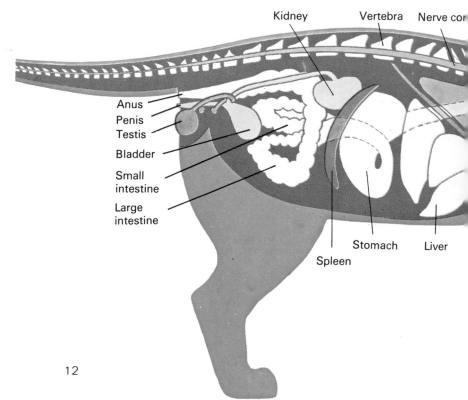

grow from the epidermis, such as hair, claws and horns, also consist of keratin.

A mammal's body cavity, unlike that of any other vertebrate, is divided into two main parts by a sheet of muscle, the *diaphragm*. This separates the thorax, which contains the lungs and heart and is surrounded by the ribs, from the abdomen, and together with the muscles between the ribs, is responsible for changing the volume of the thorax and thus causing air to flow into and out of the lungs.

The abdomen contains the stomach and intestines, which are responsible for the process of digestion, the liver, the kidneys and the reproductive system. The kidneys remove from the blood waste products which are passed to the bladder, also situated in the abdominal cavity.

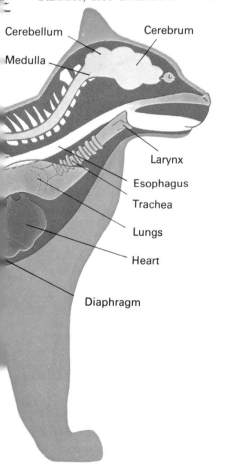

Cerebellum

Cerebrum

Medulla

Larynx

Esophagus

Trachea

Lungs

Heart

Diaphragm

The limbs

The fore and hind limbs of a mammal are built on the same basic plan. A single long bone (the humerus in the fore-limb and the femur behind) runs from the body to the elbow or knee. Next come a pair of bones side by side (the radius and ulna at the front and the tibia and fibula at the back) running down to the wrist, or carpus, which contains a number of carpal bones, or to the ankle, or tarsus, which contains tarsal bones. Next in the front foot come the meta-carpal bones, typically five in number and arranged side by side, and at the ends of these come the fingers. In the hind foot the bones in the sole are similar, but are known as metatarsals and they are followed by the toes. Although five digits, the number which mammals inherited from their reptile ancestors, are usual in land vertebrates, in the course of evolution digits sometimes disappear and many mammals therefore have less than five. A typical digit has three joints, or phalanges, but the digits corresponding to the human thumb and big toe have only two, while in the flippers of some whales extra phalanges have developed so that there may be 12 or 13.

Although the fore and hind limbs are similar in structure they have slightly different jobs to do. As the forelimb supports most of the weight it is often held straight, like

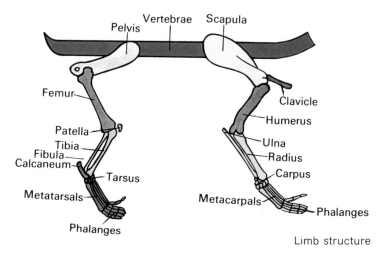

Limb structure

a pillar. It is attached to the scapula or shoulder blade which lies over the animal's back and is joined to the body by muscles which, to some extent, act as shock absorbers. The hind limb usually provides most of the power during movement, and for this reason the bends at the knee and ankle are usually clearly visible. When these bends are straightened by muscular action, a powerful kick is transmitted through the pelvis to the sacral region of the vertebral column.

The ends of the digits are protected by claws made of keratin which grow from the epidermis. Claws are useful in a number of ways. They aid in climbing, digging and scratching, and make formidable weapons. Sometimes they have become modified to form clog-like hoofs, or flattened nails.

The earliest mammals walked in a flat-footed way, with the heel making contact with the ground at every step. Mammals which have retained this condition are said to be plantigrade. A man is plantigrade when he walks, but if he runs he will rise onto his toes, thus lengthening his stride. Some mammals which are adapted for running normally move on their toes in this way—the wolf is an example— and they are described as digitigrade. Taking this trend still further, the hoofed mammals touch the ground only with their enlarged claws, thus gaining the longest possible stride. They are described as unguligrade.

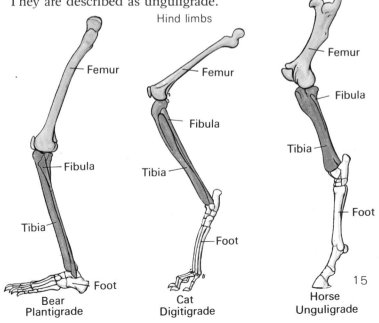

Hind limbs

Bear
Plantigrade

Cat
Digitigrade

Horse
Unguligrade

15

The importance of warm blood

The most important single factor in the success of the mammals is their warm blood. The cold-blooded fish, amphibians and reptiles have no way to maintain their body temperatures independent of their surroundings. As their bodies work best within a certain very small range of temperatures these cold-blooded animals must live in climates which suit them almost exactly, and even then they are liable to be chilled or over-heated at times. There are some compensating advantages to being cold-blooded, it is true. Reptiles expend no energy in producing body heat, and they therefore need to feed and breathe less—all energy production is work and means using food and oxygen—and they are therefore well-suited to living relatively inactive lives in warm climates.

Because they are able to maintain their own temperature at a level which is ideal for their body processes the mammals

Jerboa in dry desert

can be regarded as more 'highly tuned' than cold-blooded vertebrates are. They are thus able to maintain great activity for longer periods and, beyond this, they are to some extent independent of the climate in which they live. As long as it is not much too hot or too cold outside their bodies they are able to keep their own internal climate steady. Thus it is that some mammals are able to keep relatively cool while living in scorched deserts, while others can swim comfortably in icy seas. A single species may be found under a wide range of differing climatic conditions. For example, the puma has a range which extends from the cool Rocky Mountains of

Seal in Arctic ice

North America, across the Equator to as far south as Patagonia, where the winters can be severely cold.

No advantage is ever obtained free in this world. Because they are warm-blooded the mammals are able to be very active, but they also need to be. Heat production uses up fuel, and so food must be sought and found in large quantities. Efficient limbs, teeth, lungs, blood and so on are all essential. Also the need to maintain a constant internal temperature sets limits on the size and shape that a mammal's body can be.

Not all mammals are equally efficient at being warm-blooded. The temperature of a healthy man will always be with a degree of the normal 98.6° F, but a healthy three-toed sloth may have a temperature anywhere between 75.9° F and 99.7° F. The temperature of a bat normally fluctuates considerably during the day, reaching a maximum during active flight. Those mammals which engage in true

Zebra on African grasslands

17

hibernation lose almost all control of their temperature during their long winter sleep and are, in effect, cold-blooded at this time.

The temperature of a warm-blooded animal will almost always differ from that of its surroundings. Usually the animal will be warmer than the air or water in which it lives — this is why we speak of 'warm blood' — but it is the constancy of the animal's temperature which is the important feature. Sometimes, under tropical conditions, a warm-blooded animal will be cooler than its surroundings. In either case heat will tend to flow so as to equalize the differing temperatures. As the temperature of the animal must remain steady, the more difficult it is for heat to flow, the better it will be. This is why the shape and size of warm-blooded animals is of particular importance.

A body — living or dead — holds heat within its volume and gains or loses heat through its surface. If the flow of heat is to be kept to a minimum then the larger the volume and the smaller the surface the better. Now, the amount of surface as compared to volume depends on the shape of the body concerned. In rounded bodies the surface area is smaller in relation to the volume contained, while in elongated or flattened bodies it takes more surface to contain the same volume, and heat exchange is faster.

Because the prevention of heat exchange is important to mammals they normally have rounded bodies. Some long and thin, or flattened parts such as limbs, tails and ears may be essential, but even these are as short as possible among mammals that live in cool climates.

Size comes into it too. Large bodies naturally have larger surfaces than smaller ones of the same shape, but as size increases, the surface area does not increase as rapidly as the

Hippopotamus

volume. You can easily prove this for yourself by working out the volume and surface area of different sized cubes. Large bodies, therefore, lose or gain heat more slowly than small ones. Any of us who has eaten the smaller boiled potatoes first because they were cooler has made use of this fact. If a mammal is to be warm-blooded it must not be too small. The smallest of the shrews and mice are very near the limit in this respect.

A body in water will lose heat at a faster rate than it will in air at the same temperature. This is reflected in the spherical shape and large size of many aquatic and semi-aquatic mammals. Therefore, in view of their shape and size it is not surprising that whales can remain warm while swimming in icy seas. Large, rounded mammals like the hippopotamus have little need of hair as a warm wrapping material, particularly if they live in warm climates. In the tropics slender, elongated body-shapes are possible—the giraffe and some of the antelopes are examples. Rather small, elongated mammals from cool climates have special need of some other means of insulating against heat loss; their shape and size alone are inadequate for the task. No wonder the mink, marten and sable have such fine fur.

Marten

Antelope

Hair

Hair prevents the flow of heat from a mammal's body not because of any insulating property of its own, but because it traps a blanket of still air, which is a light but efficient insulating material. Hair consists of dead rods of keratin and grows from living cells at the bases of the hair follicles, which are ingrowths of the epidermis, the outer layer of the skin. To each follicle is attached a small muscle, the contraction of which causes the hair to stand on end. If the mammal's body is losing heat too rapidly the erection of the hairs causes the fur to trap a thicker layer of air, thus arresting heat loss. However, the elevation of the hairs beyond a certain point allows the air to circulate freely between them, and increases the rate of cooling. The sebaceous glands open into the hair follicles. They produce a greasy substance which helps to keep the hairs and the skin waterproof.

The coat of a typical mammal contains two types of hairs. Most numerous are the soft, woolly ones which make up the underfur, and which are the most important in insulation against heat loss. Growing through the underfur and overlapping on the surface so that they are the only hairs visible, there are the longer, glossier guard hairs. These hairs largely determine the color of a mammal and are often important for camouflage, and sometimes for signaling.

Hair has a number of other functions. It protects the skin from scratches and, if erected, makes its owner look larger and fiercer so as to deter enemies. The greatly exaggerated and pointed hairs of mammals like the porcupine provide a very special defense. Although hairs are dead they act as organs of touch, for pressure on them is transmitted to the sensitive nerve-endings in the dermis. Some parts of the body bear hair, such as the eyelashes and whiskers or vibrissae, which are adapted for special purposes.

Useful though hair is, it does not provide the only possible insulating layer for a mammal's skin. Fat is a poor conductor of heat, and the thick layer of blubber immediately under the skin of aquatic mammals such as seals and whales is important in preventing heat loss. It is more difficult to understand why a few land mammals, including pigs and man, should rely on fat rather than hair for insulation.

Section of skin

1 Pain sense organ
2 Lymph duct
3 Pore and sweat gland duct
4 Connective tissue
5 Sweat gland with capillary network
6 Granular layer
7 Hair follicle
8 Hair
9 Malpighian layer
10 Cornified layer
11 Sebaceous gland
12 Hair erector muscle
13 Capillary
14 Touch sense organ
15 Epidermis
16 Dermis
17 Artery
18 Nerve
19 Capillary supply to hair follicle
20 Vein
21 Pressure sense organ
22 Subcutaneous fat

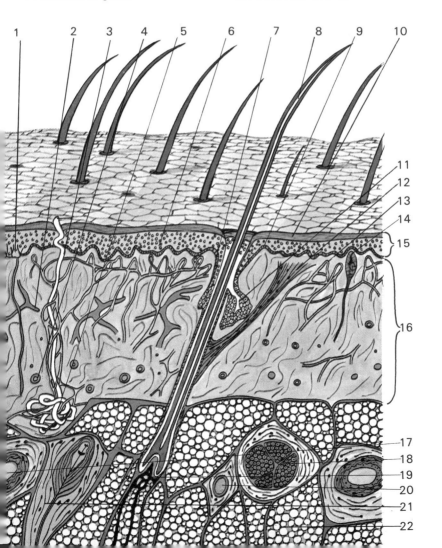

Energy

Heat is a form of energy and, like other forms of energy, it is liberated inside an animal's body as a result of a chemical reaction in which part of the food is used as fuel. Food is oxidized to produce energy, with water and carbon dioxide as waste products. Heat production is an extravagant way of using fuel —electric heating, for example, is far more costly than electric lighting— and it follows that warm-blooded animals have much higher energy requirements than cold-blooded ones. They need to eat more food, which calls for more efficient teeth and intestines, and they use far more oxygen and produce more carbon dioxide. They therefore need good lungs and an efficient means of getting the air into and out of them. This is why the mammals, unlike the cold-blooded vertebrates, have a diaphragm. Digested food, oxygen and carbon dioxide must all be carried to or from the place where energy is liberated. A good transport system, the blood, is therefore also vital.

Once produced, body heat is precious and, as we have seen, the shape and structure

Man can dissipate heat by sweating. Most furred mammals such as the wolf (*opposite above*) pant to lose heat. The Arctic fox (*opposite below*) has curled up to conserve heat.

of a mammal's body is designed to conserve it as much as possible. However, this does not mean that heat loss is always avoided. The temperature of a warm-blooded animal should be steady, not simply as high as possible. To a large extent mammals can regulate their temperatures by the postures they assume. If a mammal sprawls, with its limbs and tail extended and its tongue lolling, then cooling is more rapid. Conversely, a mammal which feels cold will curl itself up, becoming as round as possible.

Man loses heat by sweating, but this process plays a much smaller part in the lives of most other mammals because they have more hair than we do. Evaporation from a densely furred surface is bound to be slow. For most mammals, panting—and thus pumping heat out the lungs—is a normal method of cooling, while others, like elephants and rhinoceroses, bathe in mud or water to cool themselves.

Mammals tend to keep still when they are too warm, for exercise produces heat as a by-product. Chilly mammals, on the other hand, move about as much as possible.

Mammalian heart

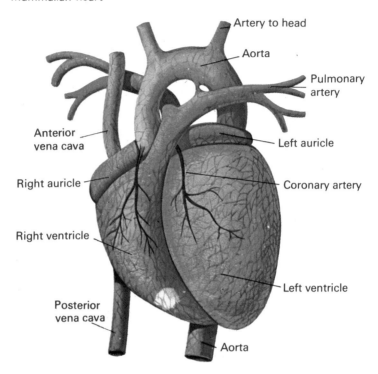

Artery to head
Aorta
Pulmonary artery
Anterior vena cava
Left auricle
Right auricle
Coronary artery
Right ventricle
Posterior vena cava
Left ventricle
Aorta

Circulation of the blood

Blood acts as a conveyor belt carrying digested food, oxygen, carbon dioxide, water and waste products. Hormones, which are in effect messages in a chemical code and are produced by special glands, are also carried by the blood. They control growth and some of the body's reactions. Substances in the blood also help to deal with any harmful bacteria which may invade the body, and seal off wounds by forming clots.

The blood passes through the lungs once every complete cycle where oxygen is absorbed and carbon dioxide is given off. The blood also passes through other tissue where it loses oxygen and picks up carbon dioxide. Thus twice during one cycle the blood passes through very narrow vessels, the capillaries. Because the capillaries slow down the rate of flow it must also be speeded up again by the

Blood circulation (diagrammatic)

Capillaries of head and fore parts

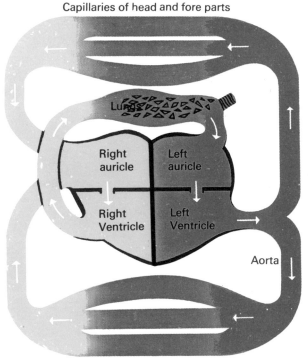

Capillaries of hind parts

heart twice during the same period. This is why mammals have what is called a double circulation. On every circuit, the blood passes through the heart twice.

The heart, which acts as a pump, is divided into completely separate halves. Its left side receives blood, which comes to it through veins from the lungs, and pumps it through arteries which lead to all other parts of the body. Having done its work there, the blood returns through veins to the right side of the heart, which pumps it through arteries which lead to the lungs. This is an efficient system and is markedly superior to the single circulation of cold-blooded vertebrates. The rate at which the heart beats varies with the sex, age and degree of activity, as well as varying from one species to another. As an example, the heart of the white whale beats 16 times per minute while the heart of the chipmunk beats 684 times.

The teeth

The teeth of a fish are specially modified scales around the mouth. Other vertebrate groups are all evolved from fish either directly or indirectly, and accordingly it is not surprising that a mammal's teeth bear a close resemblance in structure to fish scales.

Each tooth has at its center a chamber, the pulp cavity, which contains nerves and blood vessels. These enter by a channel at the base of the root. In many adult mammals the teeth stop growing, and in such cases this channel becomes very narrow. But in some mammals, such as rodents, where the paired incisors continue to grow throughout life, the channel remains quite wide. The greater part of each tooth consists of a substance very much like bone, and called dentine. The roots of a tooth are covered by

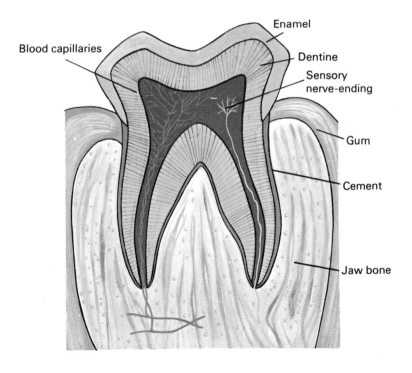

Section of molar tooth

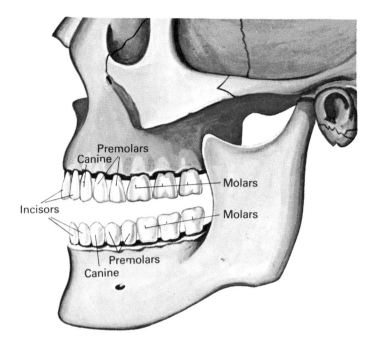

Human dentition

the alveolar membrane which holds the tooth firmly into its socket, while at the same time allowing some flexibility. This is a much stronger arrangement than a rigid joint would be. The crown of each tooth is usually covered with enamel, the hardest material to be found in a mammal's body.

Fish, amphibians and reptiles can replace a tooth in any position in the jaw several times in the course of their lives, but mammals have only two sets of teeth in a lifetime. The first, or milk, teeth are followed by more specialized adult teeth.

At the front of the mouth are the chisel-shaped incisor teeth, followed by the pointed canines, or eye teeth. Toward the back of the mouth are the cheek teeth. These consist of the premolars, which come immediately after the canines, and the molars. Molars are absent from the milk dentition, and occur only in the adult set. Cheek teeth each have two or more roots, but incisors and canines have only one each.

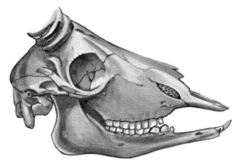

Antelope

A mammal needs a lot of food, and all of it has to be processed by the jaws before it can be digested. There is no ideal tooth which can deal equally well with all types of food, and if a mammal eats a wide assortment of foodstuffs then the design of its teeth can only be a compromise. Many mammals have become specialists, eating only one type of food, and their teeth have become specially adapted for the task. This adaptation affects both the shape and the number of teeth present.

The number of teeth which a mammal has is most conveniently expressed by the dental formula. This gives, for one side of the jaws only (as each side is usually a mirror image of the other), and for the upper and lower jaws separately (which sometimes differ), the numbers of incisors, canines, premolars and molars present in that order.

In the evolution of mammals, teeth rarely seem to have been added, although many reductions have occurred. Among the earliest known mammals the typical dental formula was:

$$I \frac{3}{3} \quad C \frac{1}{1} \quad P \frac{4}{4} \quad M \frac{3}{3}$$

Remembering that the dental formula gives only half the number of teeth, the total number present is 44. This number is seldom possessed by modern mammals, but is thought to be the ancestral formula from which the modern dental formulas are derived by reductions. For example, the dental formula for an adult human is:

$$I \frac{2}{2} \quad C \frac{1}{1} \quad P \frac{2}{2} \quad M \frac{3}{3}$$

The total number of teeth present is thus 32. Our premolars

and molars are good examples of teeth which are able to chew both plant and animal matter without being perfect for either task.

Shrew

The common shrew eats insects and other invertebrates. It has primitive, pointed teeth, which are, however, well adapted to such a diet.

The lion has well-developed canines which make useful weapons. The cheek teeth are ideal for shearing meat. There are not too many of them, for a short jaw is more powerful than a long one.

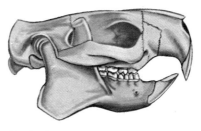

Rodent

The cheek teeth of plant-eating mammals must be able to grind up very tough plant material and they therefore need ridged surfaces. These are produced when some of the enamel wears away, exposing the softer dentine underneath. In many plant-eating animals, the upper incisors and canines are absent. Antelope have teeth of this type.

In some plant-eaters the front teeth have a more important part to play. The rodents need large, sharp incisors so that they can gnaw the harder parts of plants. Rats and mice have this type of dentition.

Lion

DIET

Although the majority of mammals are adapted to deal effectively with food of only one broad type, they are not normally narrow specialists. We know very little about the detailed feeding habits of most wild mammals, but few extremely fussy feeders are known. The koala eats only the leaves of certain species of eucalyptus trees, and the three-toed sloth eats only the shoots, buds and leaves of the cecropia plum and the hog plum trees. While we can sometimes subdivide the herbivores, or plant-eaters, into grazers which eat grasses and browsers which eat the leaves, bark and buds of bushes and trees, most herbivores, nevertheless, eat a wide variety of plants.

The same is true of the carnivores or meat-eaters. Some, such as lions and tigers, are broadly adapted for killing and eating large animals, while others, such as the small cats,

A single large meal may satisfy a lion for up to a week.

30

deal with much smaller prey. In neither case is the actual species to which the prey belongs important. Fish-eating and ant-eating mammals have equally broad tastes.

Most primates, including man, are omnivorous, eating both plant and animal matter. Some mammals that we normally think of as herbivorous are not above eating meat on occasion—red deer, for example, have been observed eating frogs.

The order Carnivora contains many primarily carnivorous species. It also contains omnivorous members, such as the brown bear. Others, such as the giant panda, are almost completely herbivorous.

Small mammals need more food in relation to their size than large ones. A shrew can eat its own weight in food in a day, but the same feat would take a lion over a month. Herbivores need a greater weight of food than carnivores because plants are not as nourishing as meat. For this reason herbivores spend more time feeding than do carnivores.

Llamas spend much of their time grazing.

31

REPRODUCTION AND GROWTH

The most primitive mammals

Nowhere is the reptilian origin of mammals more clearly revealed than in a consideration of the reproduction of the most primitive of living mammals, the monotremes. This group includes the Australian duck-billed platypus and the spiny anteaters of Australia and New Guinea. In breeding, the typical reptile lays eggs, and so do the monotremes.

Platypuses mate in water; then the female takes up residence in a special breeding burrow. Here, on a nest of leaves, she lays two eggs and in eight to ten days the eggs hatch. The young are at first incompletely formed, blind and helpless. They remain blind for nearly three months and are suckled by the mother until they are five months old.

The female spiny anteater lays only one egg, and this she carries in a pouch which develops on her body only during the breeding period. The mammary glands are situated inside the pouch, so once the young one is hatched, after an incubation period of seven to ten days, it has no immediate need to leave. It remains in the pouch for about six months until the time when its spines start to grow. It then leaves the pouch although its mother continues to look after it for some time afterward.

The female platypus builds a special breeding burrow with the entrance just above water level. When she withdraws to the nest chamber to lay the eggs, she blocks the tunnel in several places. She incubates the eggs by clutching them to her belly and, as a result, the eggs sometimes become stuck together. The incubation period is about 8 to 10 days.

The mother great gray kangaroo assists the young one to the pouch by licking a path along her fur.

The marsupials

The marsupials' name means that they are pouched animals, although not quite all marsupials have pouches. Their reproduction follows a pattern which parallels one often seen among the reptiles, where the eggs are sometimes retained inside the mother's body until the embryo has used up all the egg yolk available to it. Once this stage is reached, the young are born alive. The marsupials' evolution has followed the same path. As the amount of food available to the growing embryo inside the mother's body is severely limited, it follows that the gestation period is very short — from eight to twelve days for a dasyure (an Australian carnivorous animal about the size of a small house cat), for example, and only about 40 days for the largest kangaroos. A young great gray kangaroo is only about an inch long at birth. Its hind limbs are rudimentary bumps but the forelimbs are better developed, and using these the bright pink, naked youngster, aided by its mother who licks a path along her fur, makes its way to the pouch. Once safely inside it begins to suckle. Milk will be its only food for its first six months, and during this time it will grow considerably. By the time a young kangaroo is old enough to poke its head out of the pouch it is no longer a baby.

In marsupials where the female lacks a pouch, such as the murine opossum, the young cling to the mother's nipples.

The placental mammals

Most mammals have evolved a more efficient method of nourishing their embryos than have the monotremes and the marsupials. These more modern mammals are known as placental mammals. The embryo remains inside the mother's body during development, nourished by food carried by the mother's blood.

This is an excellent arrangement, for the embryo is provided with a good and constant environment in which to develop, and the amount of food which it can receive before birth is not limited to the amount which can be crammed into an egg. The mother is able to continue with her normal life while nourishing her young until they are born.

The number of young varies. Often, particularly with the larger mammals for which the life span is relatively long, the female normally bears only one young at a time. In man, for example, multiple births are not very common, and among elephants they are very rare indeed. In species which do not live as long multiple births are the rule. Sufficient young must be born to make up for the casualties. Where this happens each young normally develops from a separate fertilized egg, so members of the same litter resemble each other no more closely than human brothers and sisters. In some of the armadillos a single egg is fertilized and afterward divides into four separate ones, so that the female eventually gives birth to indentical quadruplets.

The larger mammals generally have the longest gestation periods. For example, the gestation period of the black rhino is 17 months and of the African elephant is 22 months. The reason is probably the relationship between surface area and volume. A large embryo will be nourished through capillaries large in area, but the volume to be nourished will be relatively larger still.

Another factor affecting the length of the gestation period is the degree of maturity which the young attain before they are born. Young rabbits, adapted to be born in a secure burrow, are blind and naked when they are born after 31 days' gestation. Young hares, fully furred and able to see, are born after a gestation of 39 days. They are better adapted for a more hazardous youth in the open.

Newborn hares—born fully furred and with eyes open.

Newborn European rabbits—born blind, naked and helpless.

The mammary glands

All mammals feed their young on milk and, indeed, it is to the possession of the milk-producing mammary glands that mammals owe their name. Milk is the perfect food for young mammals. In addition to water it contains fats and sugars, which are valuable energy-giving foods, proteins for growth, mineral salts and vitamins. The proportions of these substances present vary from one species to another. The richer the milk of a species is in fats and proteins the more rapidly the young are able to grow.

Mammary glands are formed from some of the glands of the skin. In most kinds of mammals they are actually modified sweat glands, adapted to extract the nutrients which milk contains from the blood in the capillaries of the skin. Ducts from the individual microscopic glands join and carry the milk to the nipples.

The mammary glands of the monotremes do not follow the normal pattern. It may be that they have a different origin, being modified sebaceous glands. If this is so it means that mammary glands have evolved twice during the mammals' history, and that all mammals resemble each other in possessing them not because they are all distantly related, but because the same adaptation has been evolved on separate occasions. Monotremes' yellowish milk does not emerge from nipples, but comes to the skin's surface over quite a wide area, its flow being stimulated by a rubbing action of the horny mouth of the young monotreme.

Domestic sow

The number of teats that a mammal has varies from one species to another, and is roughly related to the number of young borne by the female at one time. Where single births are the rule two functional teats are normally present, as they are in most of the primates and most of the bats. Additional rudimentary teats may also be present. The horseshoe bats have an extra pair. In this particular case they may have the special function of providing something for the young bat to cling to while its mother is flying.

Where multiple births are the rule the teats are also more numerous. Domestic pigs have seven pairs, and may exceptionally have up to 22 piglets in a single litter. Obviously no sow could raise so large a litter successfully, but the wild boar, from which the domestic pig was derived and which has not been the subject of selective breeding to make it more prolific, has a maximum litter of ten.

Teats may be situated in a mammal's chest, in the groin region or all along the under surface of the body.

Orang-utan with young, Sumatra and Borneo.

Sloth with young,
Neotropical

The care of the young

Most young mammals obtain their milk by sucking and by squeezing the nipples with their lips. This method would not work well under water, for sucking entails breathing, so the teats of whales are adapted to squirt the milk into the young's mouth. Young hippopotamuses also suckle under water, and their mothers have a similar adaptation. The quite closely related pygmy hippos are less aquatic, and in this species suckling is carried out by the more usual method.

Some young mammals are born in such a well-developed state that the period for which they are dependent upon the mother is quite short. The viscacha, a rodent from South America, is able to eat solid food less than an hour after it is born. Many hoofed mammals can run within an hour of their birth. In other cases the young are much more helpless and are dependent on their mother for much longer periods. It is difficult to set a maximum. Who would really like to name an age at which a human child becomes independent? Independence often occurs

Young animals at play. Play is thought to assist the learning processes and to develop social responses.

gradually. Be that as it may, for a period—long or short—all female mammals, occasionally assisted by the males, look after their young.

Like so many other features of mammals, parental care is connected with their possession of warm blood. Most young fish, amphibians and reptiles are left by their parents to fend for themselves. Being cold-blooded they do not need regular supplies of food, for they do not constantly expend their energy in keeping warm. For a young mammal a regular food supply is essential, and this is one reason why the habit of parental care has evolved.

While they are growing up many mammals indulge in play. This consists of activities which often closely resemble those carried out by adults, but with different degrees of intensity. Young carnivores engage in mock hunting and fighting, while young hoofed mammals sometimes skip and run away from imaginary dangers. The biological purpose of play is not at all clear. The suggestion that it is practice for the activities of adult life is an attractive one, but there is evidence to suggest that even if young mammals are prevented from playing their efficiency as adults in the acitivities concerned will not be affected. It is more likely that some kinds of play assist the learning processes which are so important to mammals, and that play with other members of the same species helps to develop the correct social responses.

Intelligence

In a sense, almost all of the behavior of most animals, and much of the behavior of mammals, is automatic. Confronted with a certain situation, the animal makes a response which is more or less appropriate, even if it has never encountered that situation before. No learning is necessary because when the animal's sense organs receive a certain set of impressions impulses flow along nerve fibers adapted for just this situation, and as a result a fixed response is produced. Behavior of this kind is called instinctive behavior.

Instinctive behavior is, by far, the most useful kind of behavior for most animals to have, for in many instances there is no time for learning. However, instinctive behavior has one great disadvantage: it is inflexible. A bird whose instincts make it feed the demanding young in its nest will continue to do so even when its own young have been ousted by a young cuckoo.

Intelligent behavior allows greater flexibility and enables animals which possess it to make more subtly varied responses to the problems they meet. Behavior of this kind is very useful, but it depends upon the existence of stored facts which have been learned. Being dependent upon their parents at the beginning of their lives, mammals have time in which to learn.

Cape hunting dogs hunt in packs and encircle their prey; Ethiopian.

40

Brain of a mammal (*top*)

Brain of a reptile (*below*)

Mammals show more intelligence than any other group of animals. The part of a mammal's brain concerned with intelligent behavior, the cerebrum or cerebral hemispheres, is much larger than that of other vertebrates. Among the mammals, intelligence varies. Whales, seals and dogs are among the most intelligent, but the monkeys, apes and—above all—man are the most intelligent of all. In this respect, if in no other, man can claim to be among the most advanced animal which has ever lived.

The nature of intelligence is well illustrated by chimpanzees which, once they have had the opportunity to play with large boxes and learn about them, are able to reach food which is beyond their normal grasp by piling boxes one on top of the other (*right*) to make a platform. Even in the most intelligent of mammals, however, much of the behavior remains instinctive, although it is often impossible to tell where instinct stops and intelligence starts.

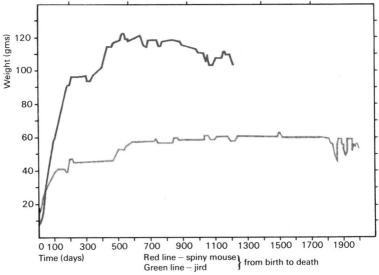

Weight (gms)

120
100
80
60
40
20

0 100   300   500   700   900   1100   1300   1500   1700   1900
Time (days)   Red line – spiny mouse⎱ from birth to death
                               Green line – jird   ⎰

Mammalian growth curve

Growth patterns in mammals

If the growth of a mammal could be measured from the moment its existence begins as a fertilized egg, it would be found that growth starts gradually, increases in rate and finally slows to a stop as full size is attained. Within this broad pattern some minor variations might well occur. For example, at the time of birth the growth rate will probably slow down until the animal has become used to its new conditions of life. In some species the onset of sexual maturity will be marked by a sudden spurt of growth. As the animal reaches old age its size might actually decrease, and the imminence of death might be indicated by some fluctuation in weight. But in essence the pattern remains. A mammal grows to full size and then stops. The length of time that growth takes depends on a number of factors, including the final size reached by the species concerned. The common shrew reaches full maturity at less than a year old, some monkeys at about seven years, and a man and an elephant, perhaps at twenty.

Growth patterns of this kind are not common to all vertebrates. Fish and reptiles usually grow, not to a fixed program, but according to the availability of food. If food is constantly available growth may persist throughout their lives. It might be said that they never reach full size, for in their case there is no such thing.

The mammalian growth pattern may, like so many other features, be associated with their possession of warm blood which, as we have seen, makes it necessary for them to have regular and frequent meals. This being so, a fixed growth pattern is a luxury which they can afford, as the raw materials essential for growth are bound to be available. This fixed program comes to an end once the animal has reached the size best adapted to the life it will lead.

Very little is known about the life span of most species of mammals in the wild. Such information as we have is fragmentary and is based partly on zoo animals which may not be typical, for they lead sheltered lives and probably often live longer than the average. The potential life spans of some mammals are:

	Years		Years
Common shrew	1½	Spotted hyena	25
Gerbil	5	Lion	30
European otter	11	Chimpanzee	40
Gray squirrel	15	Indian rhinoceros	47½
Green monkey	24	Indian elephant	70

Man is the only mammal ever known to have lived to be a hundred years old.

Shrews die the year after they are born. Elephants may live to be seventy.

THE SENSES

Active animals need keen senses, for they must be able to respond quickly to events. Naturally, each species has developed the senses most suited to its needs.

Although there are exceptions, for the mammals as a whole, olfaction, or the sense of smell, is the most important sense. Warm-blooded animals must breathe heavily, so there is a constant flow of air which can be monitored while passing through the nose. A mammal's olfactory membrane is situated inside the enlarged snout, and can be incredibly sensitive. By some means which we hardly can begin to understand it can detect quite small concentrations of chemical particles floating in the air.

Around its nostrils the typical mammal has a moist, bare *rhinarium.* Among the mammals, large, efficient noses are the rule. The only exceptions are the seals, whalebone whales, monkeys, apes and man, which have a poor sense of smell, and the toothed whales, which have no sense of smell at all.

The sense of hearing is also well developed in almost all mammals. Alone among the vertebrates they have ear-flaps, or *pinnae,* which funnel sounds into the ear opening. However, the possession of pinnae cannot be used to define the mammals, for these organs have been lost by such aquatic

The coati, with its long muzzle, has a good sense of smell; southwestern United States to South America.

Monkeys have the eyes on the front of the head to give good binocular vision. The giraffe has them at the sides to give a wider field of view.

groups as the true seals, whales and sea cows, and by some burrowing mammals. Some mammals have such keen hearing that they can gain information about obstacles in their paths by emitting sounds and detecting the echoes which are reflected. Bats and whales can do this.

The eyesight of most mammals is relatively poor. The majority are color-blind and cannot distinguish fine detail. Better vision is only found among hoofed mammals, which often live in the open and must be able to detect danger at a distance; among carnivores, which must be able to pounce on their prey with accuracy; and among the best of the climbers, for in jumping from branch to branch good vision is vital. The monkeys, apes and man have the best eyesight of all mammals, although the eyes of many birds are better still. Where the judgment of distance matters, as it does to hunters and climbers, the eyes may be located at the front of the head, but otherwise they are at the sides, for this position permits better all-around vision.

In dim light and confined spaces the sense of touch can be important. This is why most mammals have specially developed stiff hairs, the *vibrissae* or whiskers, the lips, often on the cheeks and foreheads and sometimes elsewhere.

45

Howling is a vocal signal

The blackbuck has scent-glands below its eyes.

Communication

Animals sometimes need to communicate with other members of their own species. They need a form of language, not to convey complex thoughts, but to indicate their mood and immediate intentions. In communicating with each other mammals naturally make signals which can be received by their best-developed senses. There would be little point in an almost blind species trying to use a sign language.

Often mammals have evolved special glands to produce scent. The sweat and sebaceous glands may be used, or other glands may add scent to the urine or dung, which is then used as a scent-marker with some meaning in the social life of the species. Even to our ill-developed noses mammals often seem to be remarkably odorous creatures. It is important to remember that they smell not because they are unusually dirty, but because the smell has a function, difficult though it may be for us to appreciate the subtleties of such a language.

Sound signals are often important too. As compared with smells they have the advantage of being easier to control. When it is important to remain hidden, and therefore not to communicate with another animal, sound production can stop. It is more difficult to switch off a smell. Although mammals are often silent for lengthy periods, they have voices. Even the giraffe, often believed to be dumb, makes noises on occasion. It is only to be expected that most mammals should have good voices; being warm-blooded they have plenty of breath to spare for the purpose.

Visual signals are less important to most mammals than smells and sounds, for most of them do not see clearly enough to make use of very detailed ones. Nevertheless, bold and clear visual signals are quite common. Some species have mobile tassels of contrasting black or white on their tails which make useful semaphores. The ears and other parts of the body are sometimes used in the same way. Probably only the monkeys, apes and man have sufficiently developed color vision to make use of color signals.

Making faces is a form of visual signaling much used by man. Facial signals can only be perceived with very sharp vision, and most mammals do not make them. Horses, dogs, and man's closest primate relations are among the few that do.

Lemur wafting scent signal with tail while also signaling visually

NEW WORLD

ZOOGEOGRAPHICAL REGIONS

- Palaearctic ⎫
- Nearctic   ⎬ Holarctic
- Neotropical
- Ethiopian
- Oriental
- Australasian

THE DISTRIBUTION OF MAMMALS

About a hundred years ago it was found that the distribution of birds could best be understood by dividing the world up into regions which partly, but not completely, corresponded with the continents. It was soon realized that the same regions were important in understanding the distributions of all groups of animals, including the mammals.

The zoogeographical regions represent the major landmasses which have been separated from each other to a greater or lesser extent during the last few hundred million years.

Once each region was colonized, the animals there, being

OLD WORLD

more or less isolated from others of their kind, tended to diverge from them in appearance and habits, forming new species. For example, the camel family apparently originated in North America, and from there migrated to South America (llamas) and Asia (camels proper).

The most isolated region of all is the Australasian region which has been cut off by the sea since early in the mammals' history. Sometimes part of a region has been isolated from the rest for so long that it forms a sub-region, with characteristic animals of its own. Madagascar, which is part of the Ethiopian region, is an example.

CLASSIFYING MAMMALS

As all animals are believed to have evolved from common ancestors, all of them are related to each other to a greater or lesser degree. The more closely related they are the more they tend to resemble each other. It is therefore possible to classify them according to their relationships, putting closely related species into the same small group, and this group together with others into a slightly larger one, and so on. The most important groups used for this purpose, from the largest to the smallest are: kingdom, phylum, class, order, family, genus and species. If a greater degree of refinement is required the prefixes 'super-' and 'sub-' can be used. It should be noticed that the word 'family', often loosely used to mean any related group, has a more restricted use in this system, and means a related group of a certain size only.

By no means do zoologists always agree as to how closely animals are related, and because of this they sometimes construct slightly differing classifications, joining some groups together and separating others. Disagreements of this kind can affect all groups, including species. For example, some authorities regard the reindeer of Palaearctic region as members of the same species as the caribou of the Nearctic, while others regard them as separate species. It follows that it is often quite impossible to state unequivocally how many species a group contains. For example, there are between 4,000 and 5,000 species of mammals alive today, but some authorities would quote a much higher figure.

The vertebrates form part of the phylum Chordata, and the mammals are one class within this phylum. The class Mammalia contains 18 living orders, and each of these contains one or more families. Family names always end with '-idae'. Within the space permitted by this book details can be given of most of the living families of mammals. Only in the cases of the smaller bats (sub-order Microchiroptera) and the rodents (order Rodentia) are the families so numerous that it will be necessary to group them into superfamilies, the names of which end with '-oidea', for consideration. Extinct groups are not included.

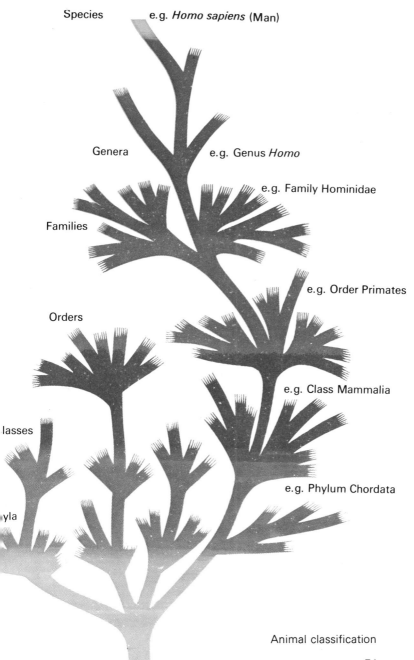

Species   e.g. *Homo sapiens* (Man)

Genera   e.g. Genus *Homo*

e.g. Family Hominidae

Families

e.g. Order Primates

Orders

e.g. Class Mammalia

lasses

e.g. Phylum Chordata

yla

Animal classification

THE MAMMALIAN ORDERS

The monotremes

Although the order Monotremata contains the most primitive living mammals, almost nothing is known of its history. The only fossils found are not very ancient and were discovered in Australia, and therefore add nothing to our knowledge of the order's distribution, for all monotremes today inhabit the Australasian region. There can be no doubt that of all living mammals the monotremes bear the strongest resemblance to the reptiles. Not only do they lay eggs, but also they have present in their skulls and shoulder girdles some bones which are more characteristic of reptiles than mammals. In certain features of their hearts and brains they resemble reptiles, as they do in having a single posterior external opening for the reproductive, urinary and alimentary systems.

The living members of the order all lack teeth when adult, although young platypuses have very small ones. All monotremes are not only primitive, but highly specialized.

The order contains two families. The family Tachyglossidae contains the spiny anteaters or echidnas. Their backs and sides are covered with spines formed from modified hairs. They have a rounded outline with short legs and broad, well-clawed feet, which are useful for digging. The jaws are long and narrow and contain a long, sticky tongue

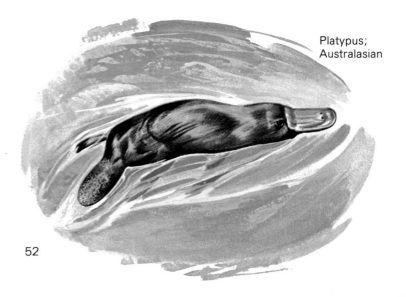

Platypus;
Australasian

Zaglossus;
New Guinea

which is used to convey ants, other insects and worms to the mouth. The two species of spiny anteaters from Australia belong to the genus *Tachyglossus,* and have shorter noses than the three species of the genus *Zaglossus* of New Guinea.

The family Ornithorhynchidae contains only one species, *Ornithorhynchus anatinus,* the duck-billed platypus, which inhabits the streams of eastern Australia and Tasmania. The platypus swims skillfully, using its flattened tail and webbed feet. It has a soft, skin-covered, leathery snout well adapted for searching the mud of stream beds for worms, shrimps, insects and tadpoles. Platypuses are mainly active at dawn and dusk, spending the rest of the time in burrows, the narrow walls of which squeeze moisture from their fur when they emerge from the water.

Like other male monotremes, the male platypus has horny spurs on his ankles. Each spur is associated with a poison gland and has a groove down which the venom can run, making these spurs formidable weapons.

Tachyglossus;
Australia

Murine opossum with young
(family Didelphidae); Mexico
and South America

The marsupials

The order Marsupialia was once widespread, for fossil remains have been found in Europe and North America, but as a result of competition from the placental mammals, its range has declined, and the order now has a discontinuous distribution. In the Australasian region the marsupials must still be regarded as successful, for until the intervention of man they met with no serious competition. A wide range of species, adapted for more modes of life than the members of any other single mammalian order, still exists there. In the Neotropical region a smaller number of species has also managed to survive, and during the last million years or so one of these managed to invade North America successfully.

Marsupials are characterized by their special type of reproduction, and almost all species have pouches, supported by the special epipubic bones. They have relatively small brains. There are eight families within the order, six Australasian and two Neotropical in distribution.

The family Didelphidae contains the American opossums, of which there are 60 to 70 species. Most of them are

Speckled Marsupial
Mouse; Australasian

omnivorous. They are all small or medium in size, usually with long, prehensile, rather bare tails. Some of them, such as the Virginia opossum, which is widely distributed in the United States, have a normal pouch, but in others it is reduced or may be entirely absent as in the murine opossum. American opposums are often arboreal, but the yapok or water opossum of South America has webbed feet, is an excellent swimmer and feeds on shrimps and fish.

The family Dasyuridae contains about 50 of the most specialized of the insect- and meat-eating marsupials of Australia. Some of them, the marsupial mice, are mouse-sized; the dasyures are the size of a small house cat and have white-spotted fur; while the Tasmanian devil is as large as a short-legged dog. The Tasmanian wolf or thylacine has quite possibly become extinct, since the last one to be seen was shot in 1930.

Little northern dasyure;
Australia

Marsupial mole;
Australia

The family Notoryctidae contains only the marsupial moles of Australia. These all belong to a single genus, *Notoryctes,* and opinion is divided as to whether there is one species or two. As their name suggests, these animals are not unlike the mole in general appearance and size, but not as a result of any relationship. They are marsupials which have become adapted for an underground life, and their resemblance to moles is the result of the separate evolution of similar adaptations. This is called convergent evolution.

Marsupial moles were discovered only 80 years ago, and because they inhabit sandy country in some of the more remote parts of Austrialia they are not often seen and much remains to be learned about them. They seem to spend less time underground than the true moles, but they burrow skillfully. They push the sandy soil out of the way with the leathery pad on the front of the nose and dig with the broad, spadelike, clawed forefeet. The eyes are tiny, hidden and useless, and the pinnae have been lost. The silky fur is whitish to golden red-brown. The pouch of the female opens not at the front, but at the back—a useful adaptation in a burrowing animal but by no means unique among marsupials. Marsupial moles feed on worms and other invertebrates.

The family Peramelidae contains the bandicoots, of which there are about 20 species inhabiting Australia and some of the other islands of the Australasian region. Bandicoots range in size from about 1 to 2½ feet long (including quite a lengthy tail). They have short legs, and the second and third toes of the hind feet are joined together for much of their length. Feet with these toes combined (syndactylous feet) also occur in the phalanger, wombat and kangaroo families, and it is thought that the claws of the combined toes form a comb useful in grooming the fur.

Bandicoots have pointed noses and large ears. Almost all of them live on the ground, sleeping in leafy nests during the day and feeding omnivorously at night. Only the rabbit bandicoots of western and central Australia dig burrows. The reproduction of members of this family is remarkable because the young do receive some nutrition through a very rudimentary placenta before they are born. They are the only marsupials to do so, although dasyures also have a suggestion of a placenta. Nevertheless, bandicoots have the typical short marsupial gestation period—about 15 days in the long-nosed bandicoot—and the young are suckled while they are in the pouch.

Long-nosed bandicoot; Australia; from 1 to 2½ ft.

57

Flying phalanger·
Australia

Rat opossum;
South America

The family Caenolestidae contains the rat opossums which inhabit the forests of South America. They seem to be rare, or at least very little is known about them, though this may be because they are shy, nocturnal and small. The head and body combined may be about four inches long, and the tail is about the same length or sometimes a little shorter. They have short legs with five toes on each foot, pointed noses and tiny eyes. In appearance, and probably in habit, they are not unlike shrews and may be adapted for the same kind of life of running in thick cover and feeding on invertebrates. Shrews are absent from all except the extreme north of South America so it is possible that the rat opossums replace them. Adult females lack pouches and probably have several young at a time.

The family Phalangeridae contains between 40 and 50 species. Its members all inhabit the Australasian region where they are often known as 'oppossums' or 'possums'. However, to avoid confusion with the American opossums and the rat opossums of the Didelphidae and Caenolestidae, it is probably safest to use the name 'phalanger' for members of the family Phalangeridae.

In size and appearance the phalangers resemble a range

of placental mammals from mice to very small bears—for the famous koala belongs to this group. Probably because they live in trees most phalangers are not unlike squirrels in appearance. Most of them have long tails which are often prehensile. They have five toes on each foot, and all of these except the big toe bear sharp claws which are useful for climbing. The big toe is opposable, which means that it works like the human thumb, and is invaluable in gripping branches. The second and third toes of the hind foot are joined for most of their length. Flying phalangers, like flying squirrels, have a fold of skin between the fore and hind limbs which enables them to make glides from trees.

Most phalangers are herbivorous, but they sometimes show omnivorous tendencies. For example the silver-gray phalanger eats buds, leaves and fruits, but also eats insects and young birds. This species is one of the most common of the Australian marsupials and has also been introduced to New Zealand, which has no native mammals of its own apart from two species of bats.

Koala;
Australia

The family Phascolomidae contains the two species of wombats of Australia and Tasmania. They are heavily built, short-tailed animals roughly three feet in length. Superficially they resemble bears even more than koalas do, but their way of life is not at all bear-like. They spend much of their life in burrows and are shy. They are often said to be nocturnal, but there is considerable doubt about this. Probably they become nocturnal when they are being closely observed. Wombats are herbivorous, feeding on bark, roots and grass. Their incisor teeth are not unlike those of rodents.

The number of toes is similar to that of the phalangers, which are probably their closest relations. However, the big toe does bear a small claw and is not opposable. A single young is born during the southern winter and remains in the mother's pouch for about six months.

The family Macropodidae—the name means big-footed ones—contains about fifty-two species of kangaroos and wallaabies. It includes the largest of marsupial herbivores, which in Australia replace the hoofed mammals of other regions. But not all of the family are large, although a fully grown great gray kangaroo standing erect may be seven feet tall. This is not quite as large as it sounds for all the family have the habit of sitting up on their hind legs. With the exception of the tree kangaroos of Northern Queensland and New Guinea, all hop bipedally with the thick, tapering tail acting as a counterbalance to the weight of the body.

Kangaroos and wallabies eat grass and leaves, often pulling the vegetation toward their mouths with their forelimbs. As far as is known only the musky rat-kangaroo (not to be confused with kangaroo rats which are rodents) is omnivorous, feeding on some invertebrates as well as plant material. Tree kangaroos spend much of their time in the trees, moving from tree to tree in great leaps. They eat leaves and fruit. Unlike most of the other kangeroos, their fore and hind limbs are nearly equal in size, but like the rest of the family, they have syndactylous hind feet, with the second and third toes almost completely joined. In all species except the musky rat-kangaroo the big toe is absent.

Tree kangaroo; Australia

Wombat; Australia and Tasmania; about 3 ft. long

Great gray kangaroo; Australia; about 7 ft. tall

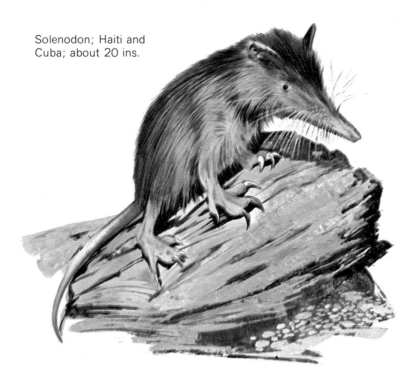

Solenodon; Haiti and
Cuba; about 20 ins.

The insectivores

The order Insectivora contains the most primitive of the
placental mammals. In structure they have changed little from
the original stock from which they, and all other placental
mammals, are descended. They are usually small animals
with pointed snouts and small brains. Their teeth are numerous and sharply pointed. The limbs are short, usually
with five clawed digits on each foot and are plantigrade.
Most of the families in the order are in decline, but a few
are still remarkably successful.

The family Solenodontidae contains only two species, one
from Haiti and one from Cuba. The solenodons are about
20 inches long including the tail. They inhabit bushy country
and forests and are mainly active at night. They are omnivorous and sometimes feed on fruit and leaves. Females may
breed twice in a year, having one or two young at a time.

Glands at the bases of the limbs produce a strong musky smell.

The family Tenrecidae is larger than Solenodontidae but has an equally limited distribution. Tenrecs are now found only in Madagascar and the nearby Comoro Islands, although fossil remains of members of the family have been found in Africa. Relatively little is known about the group and estimates of the number of living species vary between 20 and 30 species.

Some of the tenrecs, such as the spiny tenrec, have spiny hairs and bear a marked resemblance to the hedgehogs while others have remained shrew-like. One species, the rare web-footed tenrec, is aquatic and has a tail which is flattened from side to side to aid in swimming. Most tenrecs scurry about at night and are omnivorous, though some eat very little plant material. They vary from just under two feet in length down to only three inches. Some are tailless while others have long tails. Little is known of the breeding habits of most of the species, but the tailless tenrec normally has very large litters often numbering twenty or more.

Streaked tenrec; Madagascar
and Comoro Islands; from
3 ins. to 2 ft.

The family Potamogalidae contains three species of the otter shrews of tropical Africa. The largest of these, *Potamogale velox,* is one of the largest living insectivore and can reach a total length of over two feet. It inhabits rivers in the Congo region, while the two smaller species have limited distributions, one to the east and one to the northwest. Superficially they look like otters, having flattened heads, long bodies, short limbs and dense fur. The giant otter shrew's tail is flattened from side to side, but the other two species, which belong to the genus *Micropotamogale,* have rounded tails. Apart from the second and third hind toes, the feet are not webbed, but nevertheless otter shrews are superb swimmers and feed on fish and fresh-water crabs.

The family Chrysochloridae is also found only in Africa. It contains the golden moles of which there are between 15 and 20 species. As their name suggests, these animals usually have fur of a shining golden color. They look mole-like because they are adapted for the same burrowing mode of life, but their relationship to the

Otter shrew; tropical Africa; over 2 ft.

true moles is not a close one. Indeed, it has been suggested that the golden moles should be classified separately in an order of their own. Although this suggestion is not generally accepted it does indicate correctly that the golden moles stand apart from the other insectivores.

Golden moles burrow using their short, powerful limbs. There are four digits on the front feet and five on the hind ones, and each of them has a sharp, curved claw. A pad of toughened skin on the nose is used in pushing the soil out of the way. The eyes, which would be of little use in dark burrows, are tiny and are permanently closed, and the ears are small and hidden among the short fur. Although experimental proof is lacking, it must be supposed that they rely almost entirely on their sense of smell, perhaps with some assistance from the sense of touch, in finding food. Golden moles feed on worms and insects and occasionally on small burrowing vertebrates.

Cape golden mole;
Africa

Long-eared desert hedgehog;
Palearctic

The family Erinaceidae contains the hedgehogs and their relations, of which there are between 15 and 20 species. In many members of this family, such as the common hedgehog of Europe and much of Asia, some of the hairs on the back have become modified to form stiff spines. Beneath these spines the superficial dorsal muscles form a strong sheet, so that these animals can curl up into a prickly ball in the presence of danger. A few species from southern Asia, the East Indies and the Philippines have not evolved this adaptation and are known as the hairy hedgehogs or gymnures. These are probably the living survivors of an earlier stage of hedgehog evolution. The moon rat of Malaya, Sumatra and Borneo is a typical gymnure and has a more elongated shape and a longer tail than the spiny hedgehogs. It is 14 to 16 inches long and is one of the largest species of the order Insectivora.

Members of this family have five toes on their forelimbs and four or five on the hind limbs. Their snouts are pointed and their sense of hearing is usually keen. They feed on invertebrates, small vertebrates, such as frogs and rodents, and some plant material. Multiple births are common and, in the

Elephant shrew; Ethiopian;
from 7 to 23 ins.

young of the spiny hedgehogs, the spines do not appear
until after birth and are at first soft. Some hedgehogs
undergo true hibernation.

The family Macroscelidae contains about twenty species
all of which are known as elephant shrews. Of all the
Insectivora these are the most closely related to the tree
shrews or tupaias, which are sometimes considered to belong
to this order, although more often they are regarded as
belonging to the order Primates. It is probable that the
elephant shrews are primitive insectivores, similar in some
ways to the original insectivore stock from which both
modern insectivores and primates are descended. In other
aspects, however, they have become quite specialized.

Elephant shrews range from about 7 to 23 inches long,
including a tail which is nearly as long as the head and
body. They have very long, pointed noses and quite large
eyes and ears. They have longer legs than most insectivores,
with five digits on the forelimb and four or five toes at
the back. They occur only in Africa, where they live in
burrows or among thick vegetation, feeding on insects
and some leaves and fruits.

The family Soricidae contains the shrews, the most numerous and successful of all living families. Altogether there are about 250 living species, and they are found in all parts of the world except the Australasian region and parts of the Neotropical region. The family contains the smallest living mammals, the Etruscan shrew of the Mediterranean zone and the pygmy shrew of North America, both of which can weigh less than a tenth of an ounce when full grown. These tiny shrews are under two inches long. The largest species measure about 11 inches.

Etruscan shrew;
Mediterranean;
from 2 to 11 ins.

Water shrew

Shrews are short-legged with five toes on each foot. Their eyes are small and their eyesight is poor, but they hear well and their pointed noses have a very efficient sense of smell. Shrews usually lead solitary but very active lives, searching among the litter of dead leaves on the surface of the ground for insects and other invertebrates. They are fierce hunters for their size. Some species eat a certain amount of plant food as well; some species burrow; others swim well, though only the Szechuan water shrew has webbed feet. Shrews have short lives, and many species are born one year to breed and die the next. Scent glands which produce a strong musky smell are a characteristic feature of many species.

The moles and desmans, forming the family Talpidae tend to be larger than shrews, but the shrew-mole of eastern Asia, which belongs to this family, is only about six inches long including the tail and is in some ways intermediate between the two families. Desmans are long-tailed, web-footed swimmers, which feed on aquatic invertebrates, fish and amphibians, and make burrows in river banks. One species lives on the borders of southern Europe and Asia, and the other inhabits streams in Spain and Portugal.

About 15 species of moles are found in Europe, northern and central Asia and North America. They have short, broad limbs, well adapted for burrowing and each with five digits. In some species one of the wrist bones becomes elongated making it appear that there are six digits on the hand. They mainly feed on worms and insects.

Desman; Palearctic

Star-nosed mole; Neartic

69

The flying lemurs

The order Dermoptera contains only one family, the Cyno-
cephalidae, in which there are only two species. These are
the flying lemurs or colugos, *Cynocephalus variegatus* of
southeast Asia from Burma to Sumatra, Java and Borneo,
and *Cynocephalus volans* of the southern Philippines. The
name 'flying lemurs' is doubly misleading. These animals are
not lemurs (which are primates) although they do have some
superficial resemblances to them. They also cannot really
fly; they make extended leaps, using as a wing a fold of skin
supported by the shoulders, arms, ankles and tail. They
launch themselves from one tree to another and can cover
distances of over 100 yards in a single leap without losing a
great deal of height.

Adults of both species are about two feet long overall, but
are slenderly built and weigh only about three pounds. The
head is one of the most lemur-like features, for they have
large eyes and ears. The limbs are long and have broad feet,
each with five sharply clawed digits. The fur on the back is
of various shades of gray-grown and, in the case of *Cynocephalus variegatus*, is marked with white spots. Flying

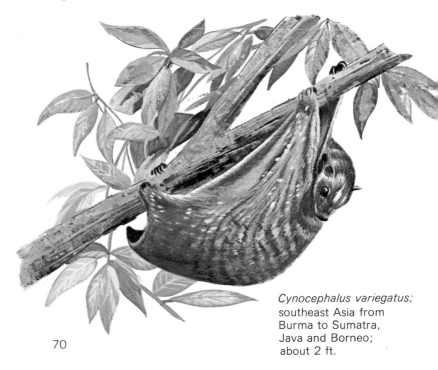

Cynocephalus variegatus;
southeast Asia from
Burma to Sumatra,
Java and Borneo;
about 2 ft.

Cynocephalus variegatus in 'flight'

lemurs are most active at night and can be very difficult to see during the day when they usually hang upside-down, flattened against the bark of a tree.

Relatively little is known of their habits, but it seems that their diet consists mainly of buds and leaves which, while hanging head downward, they pull toward the mouth with their forelimbs. They climb actively and having exhausted the food supply of one tree they leap to another. They rarely descend to the ground.

The female flying lemur normally bears a single young (although examples of twins are known) after a gestation period of about two months. Sometimes the baby is left clinging on its own while the mother is away feeding, and sometimes it accompanies her by hanging tightly on to the fur on her underside.

One of the lower incisors of the flying lemur has a number of pointed cusps which give it a comb-like appearance. The use of this structure is not certain. Perhaps it is used for grooming the fur. The true lemurs also have comb-like front teeth, but in their case each 'tooth' of the comb is in fact a separate tooth.

71

The bats

Because they are usually so inconspicuous it is easy to forget that the members of the order Chiroptera, the bats, are among the most successful of all mammals. If the number of species is any indication of success, then only the rodents are more successful. Bats occur in almost all parts of the world, being absent only from some very isolated islands and from very cold regions. There are altogether about 900 species of bats, most of which live in the tropical areas of the world.

In their basic structure the bats show considerable resemblance to the insectivores, whose ancestors they share. However, the ancestral bats branched off from the insectivore stock over sixty million years ago and, probably after passing through a phase during which they were arboreal gliders, like the flying lemurs and flying squirrels, they became adapted for powered flight with an ability which rivals that of the birds. Their forelimbs have become greatly elongated, especially the four fingers which provide such an important part of the support of the wing membrane. The thumb is also well developed, although not quite as long. The hind limb and often the tail also support the wing for, unlike feathers, the skin of a bat's wing needs to be stretched between firm supports. So that they can bend with the natural fold of the wings during flight, each of the hind limbs has become twisted outward and backward to such an extent that most bats can move only clumsily when they are not in flight. As is also the case with the birds, the large muscles of the shoulder and chest provide power for the wings. These muscles are attached to a bony keel on the breastbone.

Most bats have only one young at a time and breed only once a year. That they are able to maintain their numbers with such a low rate of reproduction is a good indication that their death rate is also low. Rodents of a comparable size breed more rapidly, producing many more young in each litter, but still only just maintaining their numbers in most situations.

The bats are divided into two sub-orders. The first of these, the Megachiroptera, contains the Old World fruit bats,

Skeleton of fruit bat

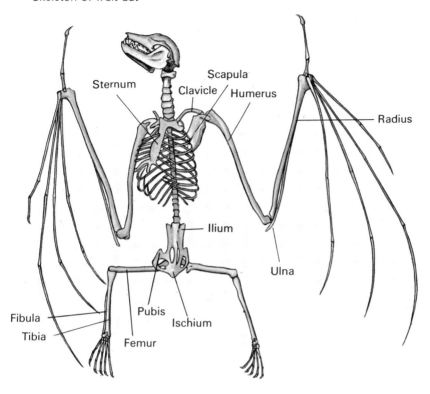

or flying foxes. There are about 150 species of them. They are often quite large, and the largest of them have wing spans of over five feet. It is hardly necessary to point out that their relationship to real foxes is very remote, but their heads are sometimes superficially like those of foxes. They have large muzzles, well-developed ears and quite large, efficient eyes. The sub-order Microchiroptera contains about 750 species, most of which are quite small. They are usually nocturnal, finding their way past obstacles not by means of their small eyes, but by reacting to the echoes of their own exceedingly high-pitched shrieks. The efficiency of their ears is remarkable. Most members of this sub-order eat insects which are caught on the wing, but some eat small vertebrates or drink the blood of large ones, and others eat nectar, pollen or fruit.

The 150 or so members of the sub-order Megachiroptera are all contained in only one family, the Pteropidae. They inhabit the tropics of Africa, Asia and Australia, and some islands of the Pacific Ocean. The largest species of flying foxes, such as some members of the African genus *Pteropus,* may weigh up to two pounds. Some of the genus *Acerodon* of Indonesia and the Philippines are nearly as large, but the pygmy fruit bat of Malaya and Sumatra, *Aethalops alecto,* weighs only half an ounce.

Flying foxes have powerful wings and often fly for considerable distances. During the day they usually roost in groups, hanging head downward in a cave or from the branches of a tree. At dusk they fly in search of food, probably mainly relying on their eyes, which are adapted to make the best use of dim light. Their keen sense of smell is also brought into play. Only the rousette fruit bats, a group with a wide distribution, and the tube-nosed fruit

Fruit bats at roost

74

Fruit bat in flight

bats make use of high-pitched sounds to guide them in the way that the Microchiroptera do. Fruit is the normal food, and dates, figs, bananas and guavas are all popular. The bat takes a fruit into its mouth and squeezes it. The pointed canine and premolar teeth pierce its skin; it is then both sucked and crushed until only a dryish pulp is left, and this the bat spits out. Although this behavior must be of considerable importance in spreading the seeds of some plants, flying foxes are not welcomed by tropical fruit growers, for the amount of damage they can do to crops is considerable. Some species also bite off and suck blossoms, probably for their nectar. The ten species of long-tongued fruit bats probably eat more flowers than fruit.

The distinction between nectar feeding and insect eating is not as wide as it may seem, for animals which visit flowers for their nectar will also find insects there. It is, therefore, not surprising that the tube-nosed fruit bats of the Australasian region should eat some insects in addition to plant material.

As their food contains much water, flying foxes probably need to drink very little, but they have been seen drinking while skimming low over rivers. The larger species of flying foxes are eaten by native peoples in some parts of the world.

The sub-order Microchiroptera contains such a great and varied number of species that they are divided into no less than 16 families. These are most easily understood if they are grouped into four super-families.

The first of these, the super-family Emballonuroidea, contains about fifty species. These include the mouse-tailed bats which have a range extending from Egypt to Sumatra and which have, as might be expected, very long tails. The sheath-tailed bats have a wider range, and have shorter tails

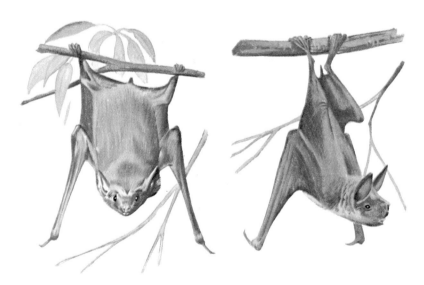

Tomb bat; Ethiopian and Oriental   Bulldog bat;
                                                Neotropical

which are loosely enveloped by the flight membrane for part of their length although the end of the tail is free. The reason for this structure is not at all clear. The sac-winged bats of tropical America, the ghost bats and the bulldog bats from the same region, and the tomb bats of the Old World also belong to this group. Tomb bats often spend the day hiding in crevices and sometimes find their way into tombs, for they are unusually agile crawlers. The bulldog bats owe

Greater horseshoe bat;
Palearctic and Oriental

Leaf-nosed bat;
Ethiopian and Oriental

their name to their faces. There are two species, and one of these regularly feeds on small fish which are seized by the bat's hind feet as it skims low over the surface of the water.

The super-family Rhinolophoidea contains about 150 species, all of which have most unusually developed noses. The skin around the nostrils bulges outward to form a grotesque structure known as the nose leaf. Exactly what function the nose leaf has is far from clear. It may be connected with the emission of the high-pitched sounds made by the bats, or it may be sensitive to very small changes in air pressure so that the presence of nearby objects can be detected in this way, but there is no real proof for either of these theories.

The slit-faced bats from the tropics of the Old World have long grooves extending backward over the face from the nostrils, but the nose leaf is relatively small. It is larger in the false vampire bats of Africa, Asia and Australia. Despite their name, these bats do not feed exclusively on blood, although they must inevitably swallow some when feeding upon their prey which includes insects, frogs, birds, rodents and other bats. In the horseshoe bats the nose leaf is very large and much folded. The part of it covering the upper lip is roughly horseshoe shaped. There are many species of horseshoe bats and their range extends from Europe to Australia. The leaf-nosed bats of the Old World have the most exaggerated nose leaves of all.

About 150 species make up the super-family Phyllostoma-
toidea, and all members of this group inhabit the warmer
parts of the Americas. Most of them belong to one family,
the Phyllostomatidae or American leaf-nosed bats, and
their common name makes it clear that they are the American equivalents of the Old World bats of the super-family
Rhinolophoidea. The nose leaves of American bats are,
however, never as large as some of those of the Old World
bats, and are in some cases very small.

American leaf-nosed bats are often quite small, but the
largest American bat, *Vampyrum spectrum,* belongs to this
group. Despite its generic name this species is not a true
vampire. As is often the case in this family, this species has
no tail, but even so the length of the head and body is nearly
six inches, and the wingspan is nearly three feet. Members of
this family roost in a variety of sites, from such typical places
as buildings and caves, to their own shelters constructed
from palm leaves. Some species roost on their own, while
others often roost with bats of other species.

The variety of food eaten by the group is unusual. Some
of them are insectivorous, some hunt larger prey including

Long-tongued bat (family Phyllostomatidae); Mexico and Neotropical

Vampire bat; Mexico and Neotropical; 6 ins. long; 3 ft. wingspan

Antillean tree bat
(family Phyllostomatidae);
Lesser Antilles

reptiles, birds and small mammals, and others feed mainly upon pollen, nectar or fruit. The pollen-feeding species play an important part in pollinating flowers.

The other family contained in this super-family has only three species, but they are of great interest. The family is the Desmodontidae, and the species concerned are the true vampires. The species are the common vampire *(Desmodus rotundus)* of Central and South America as far as Argentina, the white-winged vampire *(Diaemus youngi)* of tropical South America, and the hairy-legged vampire *(Diphylla ecaudata)* of Central and South America from Mexico to Brazil. As is well known, vampire bats feed on blood, but they do not suck it as is commonly supposed. During the hours of darkness they approach their prey stealthily—the common vampire often preys upon cattle and horses— remove a sliver of skin with their sharp incisors, and lap the blood that flows from the wound. Vampires are quite small bats and take relatively little blood, but they do considerable damage by spreading diseases such as rabies. In the course of a single year their combined efforts can cause the loss of a million cattle.

Disc-winged bat
(super-family Vespertilionoidea);
Neotropical

Little brown bat;
worldwide

The super-family Vespertilionoidea contains a number of sub-families, but most of them contain very few species. An example is the family Mystacinidae which contains a single species, the short-tailed bat *(Mystacina tuberculata)*, which is one of only two species of mammals which are native (as opposed to introduced) to New Zealand. The short-tailed bat is unusually agile when it is on the ground. It prefers forested areas, is nocturnal and is entirely insectivorous in diet.

The larger of the two most important families in this group is the Vespertilionidae, which contains over 300 species. They are small- to medium-sized bats and most of them feed upon insects, although one—the fishing bat *(Pizonyn vivsi)* of Baja, California and Sondra, Mexico—is known to catch and eat fish. They are found in most parts of the world and, except for man, have the widest distribution of any land mammal. Some species which spend the summer in temperate regions migrate to warmer areas for the winter. Like migratory birds, they have remarkable powers of navigation. For example, the little brown bat *(Myotis myotis)* has been shown to be able to return to its home when released by man from a point 165 miles away.

The other large family in this group is the Molossidae which contains about 80 species. Its members inhabit warm climates in all parts of the world. They are small-to medium-sized bats with large, peculiar growths which are used in grooming the fur. They often have a distinct musky smell, and as some of their roosts during the day may contain tens of thousands of bats, this smell can become almost overpowering. They are active at night, when they fly in search of insects. As they live in such pleasant climates they are usually active throughout the year, unlike some members of the Vespertilionidae, which hibernate. The naked bats of the genus *Cheiromeles* of Malaya, Indonesia and the Philippines are the only bats which are almost hairless.

Of the 35 species of free-tailed bats, most of which live in the tropics or subtropics, millions of members of one species *(Tadarida brasiliensis)* live in Carlsbad Caverns in New Mexico and a few other caves in the area. This species seems to prefer a diet of moths and beetles. Free-tailed bats are said to be active during most of the year.

There are two species of naked bats of the genus *Cheiromeles*. They are not in fact entirely naked but have fine hairs on the head, neck and underside of the body and tail.

The primates

As man belongs to this order we tend to think of primates as the most advanced of all mammals. This is certainly not true. The first primates evolved from early members of the order Insectivora, which is acknowledged to contain the most primitive of placental mammals. The primates, therefore, also tend to be primitive in a number of ways. For example, they still have five digits on each limb and teeth which are simple in structure when compared to those of many herbivorous mammals. In fact, the primates are best thought of as primitive placental mammals which have become adapted for a tree-climbing life. Arboreal adaptations explain most of their distinctive features, such as their dependence upon sight; their binocular (or two-eyed) vision for judging distance; their elongated fingers and toes which are adapted for gripping branches; their opposable thumbs; and, in every species except man, their opposable big toes. These digits effectively turn hands and feet into clamps.

The only primate features which cannot be explained purely and simply as a tree-living adaptations are their diet and their brains. Primates tend to be omnivorous, eating a mixture of plant and animal matter. Their cheek teeth cannot

Tree shrew; southeast Asia

be perfectly adapted for both types of food, so they are a compromise. With regard to their brains and intelligence, the smaller, more primitive surviving primates are probbably no more intelligent than most other mammals, but in the monkeys and apes the cerebral hemispheres of the brain have become large and the intelligence has become very remarkable.

The family Tupaiidae contains the tupaias or tree shrews, of which there are about 20 species. They survive only in southeast Asia. Superficially they look like squirrels, but they are definitely not rodents. Here they are treated as primates because they do have some typically primate features, such as slightly opposable big toes and thumbs.

About 15 species belong to the family Lemuridae, and they are undoubtedly primates. Although they were once widespread, for millions of years they have been confined to the island of Madagascar. They have long tails and long hind limbs, and they jump very actively. The larger species, such as the ring-tailed lemur, are cat-sized. Some are herbivorous but most are omnivorous.

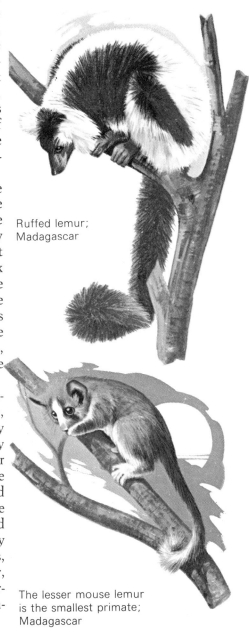

Ruffed lemur; Madagascar

The lesser mouse lemur is the smallest primate; Madagascar

Woolly indris;
Madagascar;
about 2 ft.

Aye-aye;
Madagascar

Related and similar to the family Lemuridae are two more small and rare families of primates found only in Madagascar. The first of these, the family Indriidae, contains four species. Like other primitive primates they have a dog-like moist, bare area around the nostrils, but they have shorter muzzles than members of the Lemuridae and in consequence they look more monkey-like. Their thumbs are small and the grasping ability of their hands is therefore limited, but their hind feet can grip branches very powerfully. When climbing they prefer to jump from one vertical tree trunk to another, kicking off powerfully with their long hind legs.

The woolly indris belongs to the Lemuridae group. Its head and body are about a foot long, and the tail is the same length. It is covered with woolly brown hair which is so long that it almost hides the ears. Woolly indrises are nocturnal and usually solitary. The two species of sifakas are larger. The head and body measure roughly 30 inches and the tail another 20 inches. There is considerable individual variation in the color of the long, silky fur. It ranges from yellowish-white to brown and black. A small flap of skin stretches from the arm to the body and may assist the animals in making extended gliding leaps. Sifakas are active by day, usually in groups of about six. The indris is about the same size as the sifakas, but has only a short tail. The long silky fur is usually strikingly patterned in black, gray and brown. It is active by day, but the family groups keep to thick forest and are rarely seen. All members are purely vegetarian.

The family Daubentoniidae contains only a single species, the aye-aye, which is in some respects the most specialized of all primates. Because the incisor teeth are large the aye-aye, when first discovered, was thought to be a rodent. The middle finger on its forelimb is almost incredibly long and slender. The aye-aye is about the size of a small cat and has a long tail. The dark fur is very silky. Aye-ayes are active at night, often singly. Their climbing is a mixture of laborious clambering and clumsy leaping. They feed on fruits, nuts and beetle grubs, gnawing the bark from trees and extracting the grubs with their long fingers.

Indris; Madagascar

Unusual among the primates, some of the members of the family Lorisidae are very slow climbers. These are the three species of loris from southern Asia, and the potto and the angwantibo of Africa. They are small, and the tail is either very short or absent. The fore and hind limbs are about equal in length and are usually held in a bent position, a feature often seen in good climbers.

Lorises and pottos are solitary and nocturnal. Their eyes have relatively large pupils so as to admit as much light as possible. When climbing, they move steadily, maintaining a tight hold on the branches. The smaller species, such as the slender loris, often climb among thin twigs. Only in capturing living prey, when they pounce over the last few inches, do they move with any rapidity. They feed mainly upon insects and small vertebrates, but also eat some fruits and leaves.

The six or so species of bushbabies or galagos are also usually placed in this family, although some experts think that they are different enough to deserve a family of their own. They vary in size from that of a large mouse to that of a small cat, and occur only in Africa. They have long hind limbs and long tails, and they jump very actively. If they descend to the ground they sometimes hop along on their

Slender loris;
southern Asia

Tarsier;
Sumatra, Borneo,
Celebes and the
Philippines

Potto;
Africa

back legs like little kangaroos. By day several bushbabies
may shelter together in the same hollow tree, but at night,
when they are active, they go their own individual ways.
They are omnivorous. Like all primates, the females have
relatively small litters—one or two young, and only occasionally three. The newborn young cling tightly and are carried by the mother. Again, this is a typical primate feature.

At first sight the tarsiers are not unlike wider-eyed bushbabies, but they are quite distinct, and the three species are
placed in a family of their own, the Tarsiidae. They inhabit
Sumatra, Borneo, Celebes and the Philippines. Although
like most of the more primitive primates they are nocturnal
and solitary, they have some features which hint at those
of the higher primates. For example, unlike all of the other
primates we have considered up until now, the tarsiers have
dry noses. Because their bulging eyes cannot easily swivel
in their sockets, tarsiers have very mobile necks and can
turn their heads through nearly a full circle. Insects form
the most important part of their diet.

Only two families of non-human primates are native to the New World. The family Cebidae contains the American true monkeys. Being medium-sized to large for climbing mammals, they climb by clinging with their fingers and toes, and all of their claws have become flattened nails which support the sensitive touch pads on the ends of the digits. Smaller, more primitive primates have claws on most digits. As usual in this order, American monkeys have powerful, thumb-like big toes and can grasp with all four limbs. Some of the family, the howler, spider and woolly monkeys, and the capuchins, can also cling with their tails—the only primates with this ability. American monkeys have flat noses, with slit-like nostrils displaced toward the sides.

All American monkeys, of which there are nearly 40 species, are inhabitants of warm and tropical forests, and all species except one are active by day. The exception is the douroucouli, or night ape, which is nocturnal. All the species are social, living in family groups or larger troops sometimes intermingled with members of other species. Most common in the wild are the howler monkeys, which defend a territory vocally, intimidating their neighbors with loud cries as many birds do. The capuchins and squirrel monkeys which

Woolly monkey;
Neotropical

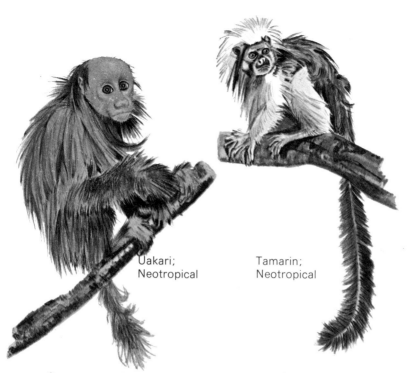

Uakari;
Neotropical

Tamarin;
Neotropical

are often seen in zoos are quite conventional in appearance, but some of the American monkeys look bizarre. For example, the uakaris have short tails and bald heads. Diets among members of this family vary. Some are truly omnivorous, some feed on leaves and some feed mainly on insects.

The marmosets and their allies are the closest living relations the American monkeys have, and they have some features which we associate with monkeys—their faces, for example. However, all of their digits except the big toe bear claws rather than nails. Admittedly the claws are flattened and in some ways a little like nails, but they are true claws. Perhaps they retain claws because their climbing is squirrel-like, depending upon claws rather than finger-pads for grip. There are about thirty species, differing considerably in color. They are sometimes divided into marmosets, which have ear-tufts, and tamarins, which lack these tufts although they may have others, and have longer canine teeth. They are omnivorous, and some species are especially fond of insects.

The Old World monkeys form the family Cercopithecidae. In a way they are the most successful primate family, for there are over 50 species. They are found only in Africa and Asia, unless some of the ancestors of the Barbary apes of Gibraltar got there of their own accord. Superficially the Old World monkeys are not unlike some of the American monkeys, but they are not thought to be very closely related. Although some Old World monkeys have small noses, some of them, such as the mangabeys, the macaques, and above all the baboons, have jutting, almost dog-like muzzles. They never have prehensile tails, although the tails of the most arboreal species are long and are of great importance in balancing on branches. The dental formula, too, is different from those of both American families because there are never more than two premolars in each part of the jaws. The Old World monkeys have the same dental formula as the great apes and man.

All members of this family are social and active by day,

Mandrill;
West Africa

Diana monkey;
Ethiopian

Male proboscis monkey; Borneo

but they are adapted for a variety of modes of life, and this is why there are so many species. Taking the omnivorous monkeys of Africa as an example, there is a range of superb climbers among the guenons (genus *Cercopithecus*) and mangabeys (genus *Cercocebus*). These are adapted for living at various levels in the forest canopy. Also in the forests, but primarily on the ground, live the drill and mandrill. In the more open areas of bush and on rocky hillsides, live the various species of true baboons. Patas monkeys also prefer the more open areas. In Asia, the macaques tend to replace the baboons as the chief ground-dwelling monkeys, although they can also climb very well. All these monkeys can store food in their cheek pouches.

A separate sub-family contains the vegetarian leaf-eating monkeys. These lack cheek pouches and have complicated stomachs which are able to digest cellulose, an important plant material. In Africa, this group is represented by the colobus monkeys which have beautiful long fur in striking patterns of white and black or reddish-brown. They are superb climbers, as are the more numerous members of this group to be found in Asia, such as the langurs and the proboscis monkey of Borneo.

(*Left to right*) Lar gibbon, chimpanzee, orang-utan, gorilla

The apes form the family Pongidae. They have no tails, but some members are nevertheless superb climbers. Some, particularly the gibbons, move by swinging beneath the branches by means of their arms, which are very powerful and longer than their legs.

The best climbers in this group are the gibbons, of which there are seven species. Gibbons are found only in the forests of southeast Asia and are the only primates which seem to be completely monogamous. A pair of gibbons will share their territory only with their own immature young. Would-be intruders are deterred by loud, whooping calls. Their diet is mainly vegetarian, but it includes some animal matter.

There are four species of great apes. Orang-utans live in the forests of Sumatra and Borneo. They live in small family groups and spend most of their time in the trees. The chimpanzees of Africa, which are slightly smaller in size, spend quite a lot of their time on the ground. They live in troops, and although they keep to their own familiar area of forest, they do not defend it against their neighbors. There

is one species of chimpanzee which includes several races. The gorillas of central Africa have the same adaptations for climbing as most other primates—binocular vision, grasping hands and feet and so on, but they have become too large to be very successful climbers. A full-grown male may weigh over 400 pounds. As far as is known, wild gorillas are purely vegetarian. There are two species of gorilla—the mountain gorilla and the lowland gorilla.

It must not be forgotten that man is a mammal, and if we arrange the mammals in a logical order it is here that he makes his appearance. His big toe is no longer opposable or thumblike, because he is a relatively heavy primate which has become adapted to walk on the ground on his hind legs, leaving his forelimbs free for other tasks. However, he still has a climber's hands and a climber's senses. He is the sole surviving member of the family Hominidae. In terms of numbers of species this cannot be judged a very successful family, but in terms of number of individuals it is very successful—even dangerously so. The apes mentioned here are now quite rare, with the orang-utan in danger of extinction.

The edentates

The name of the order Edentata means 'toothless ones', and it is inaccurate. Of the living members of the order, only the anteaters lack teeth. At the other extreme, the giant armadillo, another member of the order, has more teeth than any other mammal except some of the whales, although admittedly the armadillo's teeth are small and peg-like. This order evolved in South America, and almost all of its surviving members still remain there. In the past some of them crossed the Isthmus of Panama and invaded North America, but today only the nine-banded armadillo is found as far north as southern parts of the United States. Edentates usually have five toes on their hind feet, but often there are fewer digits on the forelimbs. Two or three of these are usually enlarged and armed with powerful claws.

The family Myrmecophagidae contains the only truly toothless edentates, the anteaters. Their skulls are greatly elongated, but contain small brains. The claw on the middle digit of their forelimbs is particularly large. There are three or four species, all of which inhabit central and northern South America. The giant anteater prefers open forests and grasslands. It lives on the ground, and like all of

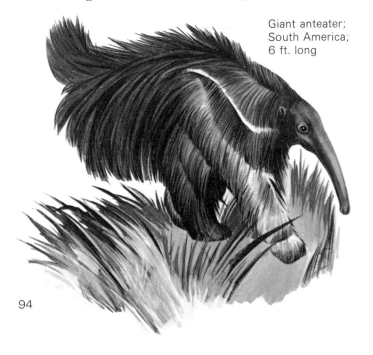

Giant anteater;
South America;
6 ft. long

94

Two-toed sloth

Dwarf anteater; South America; 15 ins. long

the family feeds solely upon termites and ants. Ant hills are ripped open with its claws, and the insects are mopped up and carried to the mouth by its long, sticky, cylindrical tongue. Including the bushy tail it measures about six feet long. The other species are all arboreal and have prehensile tails. They live in thick forests. The dwarf anteater is about 15 inches long.

The sloths make up the family Bradypodidae. There are seven species, divided into two genera. All of them are forest dwellers and can move only with difficulty on the ground. In the trees they climb slowly, hooking themselves beneath the branches with their long, curved claws. All of them have three toes on their hind feet. The sloths of the genus *Bradypus* also have three digits on the forelimbs. They feed on leaves and buds and are very fussy, preferring those of the *Cecropia* tree. The five species of the genus *Choloepus* have only two digits on the forelimbs. Almost all mammals have seven cervical (neck) vertebrae, but Hoffman's two-toed sloth has only six. Three-toed sloths usually have nine. Two-toed sloths are vegetarian but eat a wider range of plants than the three-toed sloths.

Both anteaters and sloths have only one young at a time. It clings to the mother as she moves about.

Giant armadillo;
South America;
5 ft. long

The family Dasypodidae contains about 20 species of armadillos. The skin on the back and sides of their bodies is thickened to produce a horny armor, which gives them a scaly, almost reptilian appearance. The skin bears a sparse coat of bristly hairs. They have no front teeth, and their cheek teeth are so reduced in size and structure that it is impossible to distinguish between molars and premolars. Most of them have about 30 teeth altogether, but the giant armadillo may have up to 100. The teeth are small and lack the mammalian tooth's typical hard outside layer of enamel, but they keep growing throughout the life of the armadillo.

The smallest of the family is the hairy armadillo of central Argentina, which is about six inches long including the short tail. At the other end of the scale, the giant armadillo of eastern South America measures up to five feet overall, and may weigh 130 pounds. In this species the tail is very thick and heavy, and it serves as a counterbalance to the weight of

Three-banded armadillo;
Neotropical

The armadillo protects
itself by rolling into a ball

the front of the body when the giant armadillo walks biped-ally on its hind legs. Armadillos feed on insects and other small animals and also some plant matter. In some areas the nine-banded armadillo is regarded as a pest by farmers. Some species are mainly active by day and others by night. They are all expert diggers, and they construct large burrows for themselves.

This habit of burrowing, together with their armor, provides their main defense against predators. Where the armor only protects part of the body, as it does in the hairy armadillo, this armored shield is used to plug the mouth of the burrow. The three-banded armadillos roll themselves into a ball when threatened, and their armored plates fit together very neatly during this process. The pichi, or little armadillo, of central and southern Argentina has a carapace which protects its head, back and sides. If this species is caught in the open, the legs are bent so that the unprotected undersurface is pressed against the ground.

Hairy armadillo; Argentina;
6 ins. long

Malayan pangolin;
Oriental

The pangolins

The order Pholidota has only one family, the Manidae, which contains the seven living species of pangolins, or scaly anteaters. Fossil records give no clear indication of their ancestry, and at one time they were thought to be edentates. The remarkable scales which cover the top of the head, the back and sides of the body and tail, and the outsides of the legs are not really comparable to those of reptiles. They are formed from highly modified hairs. Ordinary hairs grow between the scales and on the lower surface of the body. Pangolins have a narrow conical skull which only has room for a small brain. The mouth is small and teeth are entirely absent. The tongue is long, slender and cylindrical, and it has muscular attachments which reach back to the hipbones. There are five digits on each limb, each having powerful claws.

Some species of pangolins are terrestrial and others are arboreal, the latter species having relatively longer, prehensile

Tree pangolin;
Ethiopian

tails. They are found in the Ethiopian and Oriental regions. The four African species, the cape, giant, tree and long-tailed tree pangolins, have no external ear-flaps (pinnae) at all; but the three Asian species, the Indian, Chinese and Malayan pangolins have small ones. The giant pangolin may be over five feet long; the smaller species are half this size.

Pangolins are nocturnal and are usually solitary creatures. They seek out ants' and termites' nests by scent and, ripping them open with their claws, feed upon the occupants. During this process, their small eyes are protected from the insects by thick lids. The formic acid which ants produce as a defense is not a deterrent to a pangolin; indeed it could not digest its food properly if this acid were absent. The food is not chewed in the pangolin's mouth but passes straight to the stomach, which has muscular walls and usually contains small stones. The grinding action of the stomach is comparable to that of a bird's gizzard. Pangolins usually have a single offspring. At birth the scales are soft, but they soon harden. The young clings to the mother's back.

The lagomorphs

This order, while resembling rodents in many ways, is distinguished by the additional pair of small upper incisors situated behind the normal pair. All lagomorphs are herbivorous and, as is often the case with vegetarian animals, they have no canine teeth. The premolars and molars are packed together to provide large, ridged surfaces between which tough plant material can be ground up efficiently. The teeth continue to grow throughout life and are kept at a constant size by the wear they receive.

There are two families of lagomorphs. The first, the Ochotonidae, contains the pikas, of which there are over 12 species in Asia and North America. They are about six inches long and have short legs and no visible tail. Their ears are quite large but not as elongated as those of rabbits. They are burrowing animals and live in groups, usually in open country or among rocky outcrops. They are active by day, and have the remarkable habit of biting off vegetation,

Rabbit skull (*left*)

Arctic hares (*right*)

Pikas (*below*);
Asia
and North America;
6 ins. long

drying it in the sun and taking it into their burrows as a winter store of hay.

The family Leporidae contains about 50 species of rabbits and hares and are found in almost all parts of the world. In general the hares are larger than the rabbits and tend to live entirely on the surface of the ground, escaping their enemies by sheer speed. The jack rabbit, which is really a hare and not a true rabbit, can attain speeds of over 40 m.p.h. Smaller species, such as the cottontail, tend to hide in thick cover or construct burrows.

The Arctic hare has a white coat all year round in the far north, but in central Canada its coat turns dark during the summer. The reproductive efficiency of the rabbit is proverbial, but in fact not all of the young conceived by the female are born. At least 60 percent of them die after about 12 days' development, and are re-absorbed into the mother's body. Very often whole litters are lost in this way.

The rodents

The order Rodentia is the most successful of the mammalian orders. In terms of numbers of individuals they must far exceed all the other mammals put together. There are about 1,800 species of them. The reasons for their success are clear enough. They are small—the largest is no bigger than a small sheep—and they are basically herbivorous. Herbivores are always more numerous than carnivores, for plants are easier to come by than meat. They can be burrowers, runners, swimmers, climbers or gliders, and numerous rodents excel in all of these activities in most parts of the world.

Rodents always have one persistently growing incisor on each side in both the upper and lower jaws, and never have canines. The cheek teeth have ridged surfaces and are adapted for grinding plant matter. The muscles which move the jaws are very large, and on the basis of the arrangement of these muscles, together with certain other features, the rodents are divided into three sub-orders. These are the Sciuromorpha (squirrels), Myomorpha (rats and mice) and Hystricomorpha (porcupines and many other rodents, most of which are South American). Within these sub-orders are a number of super-families, and it is at this level

Mountain beaver—not a true beaver; western North America

European red squirrel; Nearctic

that the group will be dealt with here.

There are five super-families of sciuromorph rodents. The first, the Aplodontoidea, contains a single species, the mountain beaver of western North America. Despite their name, these animals are burrowers rather than swimmers, though they are often found near rivers.

The super-family Sciuroidea contains about 250 species. These are the squirrels, flying squirrels, marmots and chipmunks, and members of this group inhabit all parts of the world except the Australasian region. Many of them are long-tailed successful climbers, and the flying squirrels have made gliding a natural extension of this activity. However, the majority of species spend most of their time on the ground, as the little chipmunks of the Holarctic region do. The marmots are short-tailed, burrowing animals.

The super-family Geomyoidea is confined to North and Central America, where there are over 100 species. These include the small, burrowing pocket gophers, the pocket mice and spiny pocket mice, and the desert-loving kangaroo rats.

Prairie dogs;
Nearctic

Eastern flying squirrel;
Nearctic

Pocket gopher;
Nearctic

103

Beaver; Holarctic; 3½ ft. long

The sciuromorph super-family Castoroidea contains only the true beaver. They average nearly 3½ feet in length, and inhabit North America and parts of Europe and Asia. There are five toes on each foot, and those of the hind feet are webbed.

Beavers live in family groups, made up of one pair and their young. They live in dens which open beneath the surface of still, fresh water. To this end, beavers build dams across rivers. The largest dams known are over a quarter of a mile long. Beavers feed mainly on the bark of such trees as willow and alder, although in summer they also eat

Scaly-tailed squirrel;
Ethiopian

Springhaas;
Ethiopian;
about 2½ ft.

waterweed, cattails and other aquatic plants. The female usually has from two to four young, and they stay with their parents until they are at least two years old. Beavers may live for over 20 years.

The super-family Anomaluroidea contains nine species. Most of these are scaly-tailed squirrels, which owe their name to the possession of two rows of scales made of thickened skin on the underside of the base of the tail. These enable the tail to obtain a grip on the bark of trees, although it is not in any way prehensile. These animals range in size from that of a large mouse to that of a squirrel, and are found only in the forests of Africa. All of them except one species have gliding membranes very much like those of the flying squirrels. The exception is *Zenkerella insignis* of West Africa.

The only other member of this super-family is also found only in Africa. It is the springhaas or springhare, which looks like a cross between a kangaroo and a rabbit. Including the long tail, the springhaas is about 30 inches long. Despite their lanky appearance, they are expert burrowers. They usually live in pairs and emerge from shelter at night in order to feed on roots and other plant material, including crops. They sometimes travel for long distances and can move at considerable speed.

Lemming; Holarctic   Water vole; Palearctic

The rodent sub-order Myomorpha contains a large number of species, and most of these belong to the super-family Muroidea. There are perhaps 1,200 species in this super-family altogether, and they are very widely distributed. These animals never have premolar teeth. Many of them have long tails, although some do not, and they are never very large in size. Most of them burrow and live on the ground, but some are arboreal and others are aquatic. Like many other rodents, a lot of them store food. As they are largely vegetarian this a valuable adaptation, for many kinds of plant material keep well and are often only seasonal.

Among the better-known species are the hamsters of Europe and central Asia. These have very short tails, brown fur on their backs, black fur underneath and white markings on their sides. They live in burrows and in some areas are regarded as a menace to crops. They are related to the golden hamster which, although it is now well known as a pet, was only discovered in 1930. All of the countless members of this species kept in homes and laboratories are descended from a single litter, and next to nothing is known about this species in the wild. Lemmings are also short-tailed. The remarkable migrations of one species, the Norway lemming are accompanied by a rapid build-up of the population, which is followed by an even faster decline in numbers. These cyclic variations in population are characteristic of a number of members of this group, including some of the voles. The rapid rate of reproduction encourages an increase in numbers until the population tends to outrun its food supply. Other factors are at work too. When the rodents become numerous so do their predators, and perhaps large rodent populations are unusually susceptible to some forms of disease.

Rats and mice have become associated with men in a way which must be as old as human civilization. The house mouse originally came from southern Europe and north Africa and the black rat from southern Asia. Both have accidentally been carried to all parts of the world by man. The brown rat originated in eastern Asia, and its spread has been more recent than those of the other two species. It has only reached some remote parts of the world during the twentieth century. Apart from the damage which these species do to human food and property, the rats are also the carriers of diseases such as typhus and bubonic plague, and their effect on history has been incalculable.

The mole rats of southern Europe, Asia and northern and eastern Africa, and the bamboo rats of southeast Asia stand apart from the rest. They are stoutly built animals and, as their names suggest, are either adapted for an underground life, or spend much of their time hiding among bamboo thickets.

Brown rat; worldwide

African mole rat; Ethiopian

About thirty species of myomorph rodents belong to the super-family Gliroidea. These are the dormice. They range in length from four inches to just over a foot long, including a long, usually bushy tail. Most of them are good climbers and look like small squirrels. They are usually nocturnal and omnivorous. They inhabit Europe, Asia and Africa.

The best-known example is the common dormouse which inhabits the deciduous woodlands of Europe. It prefers areas in which the undergrowth is thick, especially if honeysuckle is present. At night the common dormouse climbs actively, grasping slender branches with its toes. If feeds on seeds, fruit and nuts and perhaps occasionally on insects. It does not store food, but with the onset of winter it goes into true hibernation in a well-protected nest below the ground or in a hollow tree. A female may have two litters in a year, usually numbering three or four young ones. The largest of the dormice is the fat dormouse, or edible dormouse, which originally came from central and southern Europe, but which now has a larger range in Europe and has been introduced into Great Britain.

The spiny dormice are inhabitants of India and southeast Asia. They have little hair on their tails except for a tuft at the end. Spines are scattered among the hair on their backs.

The super-family Dipoidea contains about 35 species of birch mice, jumping mice and jerboas. All of them are adapted for jumping and have long, powerful hind limbs. The birch mice inhabit Europe and Asia, living in forests or on open steppes. They are usually nocturnal. In winter they hibernate. Birch mice produce no more than one litter each year, and individual females bear only two litters within their lifetimes. Apart from one Chinese species, the jumping mice all come from North America. On the whole they live on the ground, and can cover six feet at a single bound if they are startled. The jerboas, which are larger, can jump over nine feet. They inhabit dry areas of central and southern Asia and North Africa. Most of them have only three toes on the hind foot. They are nocturnal and feed on cactus, other succulent plants and seeds. As far as is known they obtain sufficient water for their needs from their food and do not drink in the wild.

Fat dormouse (*top left*)
Common dormouse (*top right*)
Birch mouse (*left*)
Jerboa (*below*)

Old World porcupine *Hystrix*;
Ethiopian

Brush-tailed porcupine
(Old World); Oriental

North American porcupine;
Nearctic

In their external appearance the members of the sub-order Hystricomorpha are the most varied of the rodents. They are divided into no less than seven super-families.

The super-family Hystricoidea contains the porcupines of the Old World, of which there are about fifteen species. They are found in warm areas from Italy and Africa to the Philippines. They have five digits on each foot and are, for rodents, quite large, some being about three feet long. The spines covering the body are stiff, sharp hairs and may measure over a foot in length. When the porcupine is frightened, its spines stand on end and provide an efficient defense. They are expert diggers, and construct deep burrows for themselves. A number of porcupines may share the same burrow. They feed on bark, roots, fruit and foilage.

The New World porcupines are classified in a different super-family, the Erethizontoidea. This is because they are thought to have evolved the features which they share with the Old World porcupines, such as their spines, quite inde-

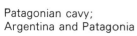
Patagonian cavy;
Argentina and Patagonia

Capybara;
Neotropical

pendently of them. There are about ten species of American porcupines. Again, they are relatively large rodents, but they differ from the Old World porcupines in having broad hind feet, especially adapted for climbing trees. They have only four working digits on each foot. They are predominantly tree-dwellers, and most of them have prehensile tails. The North American porcupines lack the prehensile tail of many in this group. They eat the bark of trees and, for this reason, are considered pests in some areas.

The super-family Cavoidea has representatives only in central and South America. For rodents they are medium-sized to large animals, and they all either burrow or run on the surface of the ground. There are about 35 species. They include the cavies—the wild relations of the guinea pig—which burrow or hide among rocks, the long-legged Patagonian cavy, or mara, which runs in the open in small herds; and the largest of the rodents, the capybara, which is as large as a small sheep. These inhabit thick jungles, and are never found far from water in which they take refuge when frightened. The paçarana and paca also live in forests, and they look like large, spotted guinea pigs. The more slender agoutis and the smaller acouchis have very lightly constructed limbs and usually escape danger by running, although they sometimes hide in burrows.

Viscacha; Argentina; over 2 ft. long

Coypu, or nutria; Neotropical

Another of the super-families of hystricomorph rodents is best known for the fur borne by some of its members. This is the super-family Chinchilloidea. It contains only six species, all of which have only either three or four toes on the hind foot and large ears. They all have long tails and inhabit South America. The viscacha is a creature of the open grasslands and bush. The head and body are about 20 inches long, and the tail another seven inches. They live in colonies, each animal occupying an extensive system of burrows, the openings of which are marked by heaps of the stones which have been unearthed during burrowing operations. They are herbivorous. The four species of mountain viscachas are a little smaller, but they have longer tails. They live near streams in the foothills of the Andes. The chinchillas are smaller still, having a head and body about ten inches long and a tail five inches long. They also prefer rocky hillsides and are herbivorous. At one time they were very heavily hunted for their fur, but the species is now protected and most of the chinchilla pelts used in the fur trade come from animals bred on fur farms.

The super-family Octodontoidea contains about 100 rat-like species, most of which are American. However, a few members of this group are natives of Africa. The best-known member of the super-family is the Coypu, a species which originated in Central and South America, but which has become naturalized in the southern United States and some other parts of the world as a result of individuals escaping from fur farms. Coypu fur is not especially valuable, but it has soft under-fur beneath the stiffer, glossy guard hairs and is known in the fur trade as nutria. Wild coypus live along the banks of streams, where they construct burrows. They are good swimmers and have webbed hind feet. They breed rapidly and may have as many as nine young in a litter. In the South they are regarded as pests because of the damage they do to crops, and because of the harm they can do to drainage channels.

Degu; Peru and Chile

Tucotucos belong to a number of closely related species are found in a variety of habitats in almost all parts of South America. They have a rounded outline and short limbs armed with powerful claws. They are burrowers and live in colonies. They rarely venture far from the shelter of their burrows. Also contained in this super-family are the South American spiny rats, which provide yet another example of mammals with stiffened, spiny hairs. Cane rats and rock rats of Africa also belong to this group.

Tucotuco; Neotropical

Spiny rat; Neotropical

Cape mole rat;
South Africa;
about 8 ins.

The super-family Bathyergoidea is a small one and includes only one sub-family. This contains the African mole rats, which have short tails and limbs, thick-set bodies, and very small eyes and ears. Like the moles, most species have short, velvety hair and broad hands and feet which are well adapted for digging. They feed on underground roots and plant stems, and some of them also eat some invertebrates. Their incisor teeth are very large, and are usually visible even when the mouth is closed. They are sometimes used to assist the limbs in digging. The number of cheek teeth is very variable, and in some cases exceeds the number normally found in rodents. There are 15 to 20 species.

The Cape mole rat is in some ways typical of the group. It inhabits the southern tip of Africa, and its average length is about eight inches. The coloration and markings on the fur are variable. It spends most of its life underground and prefers sandy soil. The line taken by its burrows is marked by small heaps of excavated soil, like molehills. One of the burrows is widened to form a rounded chamber, and here some plant food is stored as a reserve against times of drought, when

Naked mole rat; East Africa

the ground may be too hard for digging. The most specialized species is the naked mole rat of parts of East Africa. There are some fine hairs scattered over the surface of its body, but they are few and far between. There are no external ears. Only rarely does this species come to the surface of the ground.

The only group of hystricomorph rodents remaining to be considered is the super-family Ctenodactyloidea. This is another small group, containing only one family which consists of four species, all called gundis. Although fossils of members of the group have been found in Europe and Asia, its surviving members occur only in dry parts of northern Africa. They look a little like guinea pigs but have visible short tails. At night and during the hottest part of the day they hide among rocks, coming out to feed early in the morning and in the evening. The fleshy leaves of desert plants are their favorite food and also provide them with water. If they are threatened they normally run for cover, but if cornered they play dead.

Gundi;
North Africa

The cetaceans

The order Cetacea contains the whales, a group of mammals which became aquatic tens of millions of years ago. The smallest members of the order are only just over four feet long when they are full-grown. However, the whales also include the largest animals of any kind that have ever lived. They have streamlined bodies with no visible necks. A thick layer of blubber under the skin helps the streamlining and also conserves body heat. The forelimbs have become short-ened and broadened to become fin-like, and their digits have evolved extra joints. The hind limbs have disappeared completely, and only vestiges of the hip bones remain. Unlike that of a fish, the tail of a whale is flattened horizon-tally, and during swimming it moves up and down rather than from side to side. They must surface to breathe, but can stay submerged for long periods; two hours is the record. Whales are completely aquatic and even sleep, give birth to their young and suckle them in the water.

Whales are divided into two sub-orders. The first of these, the Odontoceti, or toothed whales, always retain at least some teeth, and often have a great many although each tooth is usually conical and simple in structure. Members of this group have only a single nostril which, as is always the case in whales, is on top of the head. The other sub-order, the Mysticeti, or whalebone whales, retain two nostrils, but lack teeth, having rows of brush-like plates.

The family Platanistidae belongs to the first of these sub-orders and contains four species. These animals are

Geoffroy's dolphin;
Amazon and Orinoco
River systems;
up to 10 ft.

Beaked whales;
all oceans;
between 14 and 40 ft.

rather small, never exceeding ten feet, and inhabit rivers, estuaries and sometimes coastal waters. The jawbones are formed into a prolonged beak which is armed with many teeth—sometimes over 200 of them. They feed mainly upon invertebrates. The species are the Ganges dolphin of southern Asia the Chinese river dolphin of southeast Asia, and Geoffroy's dolphin and the La Plata dolphin, both of South America.

The family Ziphiidae contains the beaked whales, of which there are 18 species. They have a widespread oceanic distribution. In size, they range between 14 and 40 feet. They have only one or two pairs of teeth, all in the lower jaw, and feed on squid and fish.

Sperm whale;
all oceans;
from 30 to 60 ft.

The family Physeteridae contains only two species. These are the sperm whale and the pygmy sperm whale. Both of these species have small dorsal fins, but that of the sperm whale is not very obvious, being low and rounded in outline. Male sperm whales measure up to about 60 feet long, and are half as large again as the females. This species has always been one of the most important to the world's declining whaling industry. Apart from oil, it yields spermaceti, a waxy substance used in making some ointments. In the living whale, spermaceti, which is light in weight, probably helps to adjust the animal's buoyancy so that, for example, the part of the head which bears the nostril is the first part of the body to break the surface of the water.

Sperm whales range the oceans of the world. They travel in schools, some consisting of one adult male together with a number of females and their young, while others consist solely of males. They feed mainly on cuttlefish and squid, including some giant species which are obtained at depths of up to 3,000 feet. The banana-shaped teeth are borne only in the lower jaw and fit into sockets in the upper jaw when the mouth is closed. There are usually about 40 or 50 teeth altogether, but as they are all the same shape it is not possible to divide them into different kinds. This state of affairs is normal in the toothed whales. Females give birth to a single young after a gestation period of about 16 months. Sperm whales have been proved to live for 32 years, but it may be that they can live longer than this. The pygmy sperm whale is only about ten feet long, and although its body contains both oil and spermaceti it is too small to be of much interest to whalers. It inhabits the Atlantic, Pacific and Indian Oceans.

The family Monodontidae is another small one with only two species, both of which inhabit only Arctic and northern seas, preferring coastal waters. The white whale, or beluga, is dark in color when it is young, but becomes white at about five years of age. It measures up to 15 feet long and has about 36 teeth altogether, equally distributed in both upper and lower jaws. The other member of this family, the narwhal, is about the same size. It has only two teeth, both in the upper jaw. In males, one of these teeth, usually the one on the left side of the body, becomes a huge spirally grooved tusk up to eight feet long. The function of this tusk is not at all clear. Both white whales and narwhals feed on shrimps, squid and fish.

Male narwhal;
Arctic seas;
about 15 ft.

There are about 60 species of small- to medium-sized whales which are sometimes grouped together in a single family, but which are best divided into three families. The family Stenidae is distinguished from the rest by the arrangement of the sinuses of the skull. It contains eight species, which among them are found in all oceans of the world. The largest of them, the West African white dolphin, measures up to eight feet in length. The name 'dolphin', which members of the Stenidae share with most of the next family to be considered, is applied to small, graceful cetaceans which have pointed, beak-like jaws. Relatively little is known about the habits of members of the Stenidae, and some of them are rare. They often swim in large schools, but the Amazon white dolphin, the only species which is confined to fresh water, lives in groups of two or three.

The family Delphinidae contains over 40 species. Most of them have distinct beaks, and the remainder, including the grampus, and the killer and pilot whales have rounded, often bulging foreheads. They are speedy swimmers and often jump right out of the water. They do not dive to great depths. The smaller species, such as the common dolphin which is about six feet long, often swim in large schools. One school of this species was estimated to contain about 100,000 individuals. The common dolphin has a worldwide distribution and feeds on fish and squid. It may have up to 200 pointed teeth, situated in both the lower and upper jaws. The closely related cape dolphin may have over 240 teeth, while at the other end of the scale the grampus has only about ten teeth, all in the lower jaw. In relatively recent years some species of this family have been successfully kept in captivity, and have been found to be remarkably intelligent. The bottle-nosed dolphin is an example. This species, like many other cetaceans, emits high-pitched sounds and avoids obstacles by listening for the echoes in the same way that many bats do. The larger members of the family include the killer whale which is about 20 feet long. Killer whales usually inhabit Arctic and Antarctic seas. They hunt in packs of up to 50 individuals and feed on other whales, seals and penguins. They have huge appetites. The stomach of one killer whale was found to contain remains of 14 seals and 13 porpoises.

The family Phocoenidae contains seven species of small, blunt-nosed thick-set porpoises. They have a small beak. The common porpoise measures up to six feet long and swims slowly in schools of up to 20 in coastal waters and estuaries. It feeds upon fish such as herrings. Others belonging to this group, such as Dall's porpoise of the Pacific Ocean, are among the speediest of all cetaceans. Members of this family have teeth with a curved, spade-like outline. They have between 60 and 120 in all, borne equally in the lower and upper jaws.

Killer whale;
Arctic and Antarctic
seas; about 20 ft.

The three families of whalebone whales contain most of the very large whales. Despite their size they feed on very small organisms, such as small fish, shrimp and plankton, which they collect in a very efficient way. The whale gulps a great mouthful of sea water and, closing its jaws and using its tongue as a piston, forces it out again. The water is strained through the baleen (closely packed whalebone plates), and the rich food contained in the plankton remains inside the whale's mouth.

The family Eschrichtidae contains only the gray whale of the North Pacific Ocean. This species measures up to 50 feet long. At one time gray whales were hunted by whalers, but in recent years they have been protected. Unlike most large whales, they sometimes enter very shallow coastal water.

Because they alone among the whalebone whales have dorsal fins, the six species which make up the family Balaenopteridae are sometimes collectively known as finback whales. They are also known as rorquals, and can be identified by the row of grooves which run from beneath the floor of the mouth along the lower surface of the body. The rorquals are streamlined and powerful animals, and include the fastest of cetacean swimmers. The smallest of them is the lesser rorqual, or little piked whale, which measures a maximum of only 30 feet. Another member of the group, the blue whale, is the largest animal which has ever lived. They can measure nearly 100 feet long and weigh about 120 tons. This species is widely distributed but is now near extinction due to the large numbers killed by whalers. It swims alone or in small groups. The single young—the almost invariable number in whales—is about 25 feet long at birth and grows rapidly. It becomes sexually mature at about five years of age.

The family Balaenidae contains the right whales, which may owe their name to the fact that they were considered to be the right whales to catch from a whaler's point of view. There are five species which among them are found in all oceans, especially the cold seas where the plankton is particularly rich. They have huge heads in relation to the size of the rest of the body and range in size from the pygmy right whale, a mere 20 feet long, to the Greenland right whale, or bowhead, which can reach 60 feet.

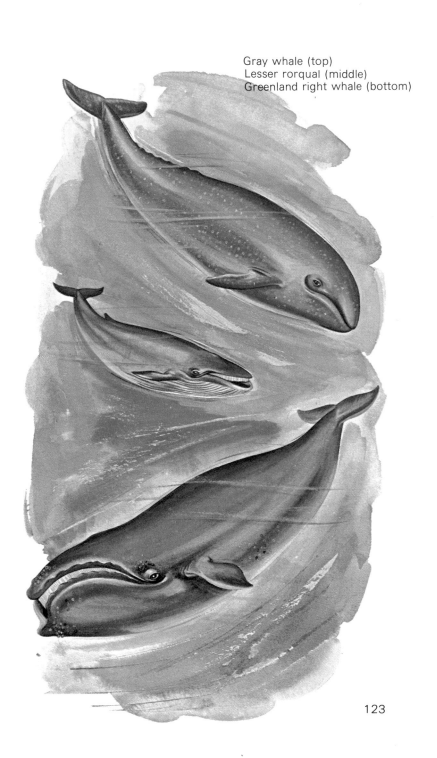

Gray whale (top)
Lesser rorqual (middle)
Greenland right whale (bottom)

123

Wolves; Holarctic
and Oriental regions

The carnivores

The order Carnivora is one of the most successful of mammalian groups. Mammals are active animals, and the Carnivora as a group are adapted for modes of life in which great activity can be invaluable; they are hunters. They typically have keen senses in order to find their prey, active bodies so that they can catch up with it, and killing weapons — especially the pointed, curving canine teeth, and cheek teeth which act as the blades of shears in cutting up flesh. Having fed on meat, which is a very filling type of food, the typical carnivore is able to indulge in a lengthy rest before the next meal. Of course, despite the name of the order and the adaptations of its members, not all of the Carnivora are purely carnivorous. Most of them eat some plant food, too. From the dietary point of view it would be more accurate to describe many of them as omnivorous. In a few cases this trend toward plant-eating has taken an exaggerated form, so that plant food has become the most important part of the diet, but even so the Carnivora as a whole are adapted as meat-eaters.

The order is divided into two sub-orders, and these are so distinct from each other that many experts prefer to regard them as two separate orders. The are the Fissipedia (or split-footed ones) in which the toes are typically separate, which are basically land animals, and the Pinnipedia (fin-footed ones) which contains the seals, sea lions and walruses, all adapted for life in water.

The sub-order Fissipedia contains seven families. The first of these is the Canidae, or dog, family which contains about 35 species. Essentially these are runners, for they are digitigrade. This means that they move on their toes. The large number of cheek teeth present means that the wild dogs have long muzzles and large noses, and it is accordingly not surprising that their sense of smell is exceptionally keen. The typical wild dog in many respects is the wolf, which hunts in packs of six or so individuals. They use their sense of smell in order to trace their prey and run tirelessly in a group until they catch up with it. Wolves are widely distributed in the Holarctic and Oriental regions, and in other regions there are other species living in very much the same way. The coyote of North America and the jackals of Africa and Asia are smaller species which rely to a larger extent on scavenging. The foxes are more solitary than other dogs, and the most unusual members of the family are the small bush dogs of South America.

The seven species of bears which make up the family Ursidae can be thought of as heavyweight, flat-footed dogs. They are not good runners, but can use their size and considerable strength in overcoming very large prey species. Apart from the polar bear, which feeds mainly upon seals, they are omnivorous. The brown bear has a Holarctic distribution, and other species are found in North America, South America and Asia.

Brown bear: Holarctic

The family Procyonidae contains the raccoons and the pandas. Typically members of this group are smaller than the dogs, have long dog-like muzzles and plantigrade limbs like the bears. Both their size and limbs adapt them for climbing, and they usually live in woods and forests, often close to rivers. They are omnivorous, and some of them show a tendency to live mainly on plant food. The raccoons inhabit a wide range of climates from North to South America, while the pandas are found only in Asia. The giant panda which lives in the bamboo forests

Raccoon; Nearctic and Neotropical

Giant panda; central China

126

of central China is the largest member of the family and looks very much like a bear and is considered by some to be one. Its diet consists almost entirely of young bamboo shoots.

The family Mustelidae contains the weasels and their relations, of which there are about 70 species. They are found in all parts of the world except the Australasian region. Although some are omnivorous, most of them are fierce hunters and because they have fewer cheek teeth than the families so far considered they have shorter jaws and very powerful bites. Most of them are small, and they tend to have long, muscular bodies and short limbs, which may be either plantigrade or digitigrade. Some, like the weasels, are well adapted for hunting in thick cover or below ground, while the martens have long tails and can climb trees. The badgers and their allies are among the heavier members of the family. Most of the group have well-developed scent glands, and these reach the peak of their development in the skunks. The otters have evolved webs of skin between their digits and have become excellent swimmers, one species even having taken to the sea. Members of this family from cool climates have superb soft, warm fur.

Mink

The sea otter has the remarkable habit of bringing
up stones from the sea bed
along with shellfish, and using
the stones to crack the shells.

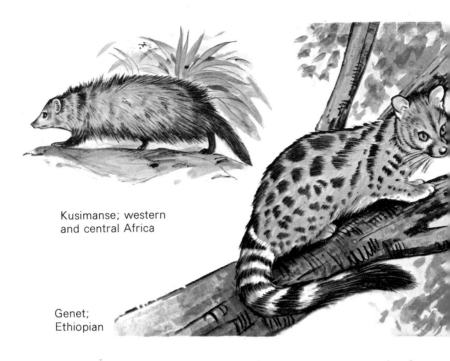

Kusimanse; western
and central Africa

Genet;
Ethiopian

The remaining families of land carnivores are more closely
related to the cats than to the dogs. About 80 species make
up the family Viverridae, the members of which resemble
the Mustelidae in many ways. Nevertheless, they are certainly separate groups, and the resemblances are due to convergent evolution—the separate evolution of similar charac-
teristics—rather than close relationship. Members of the
Viverridae occur naturally only in the Old World, including
parts of Europe, Africa and southern Asia. Mongooses have
been introduced to parts of the West Indies and New
Zealand by man. Like most introductions of this kind,
this is now seen to have been a great mistake.

Typically members of the Viverridae have short limbs
which are either plantigrade or digitigrade and bear partially
retractile claws. Including the tail they are between one foot
and about six feet long, most of them being nearer to the
lower end of this range. They are mainly carnivorous, but
some are omnivorous, and the palm civets feed chiefly on
fruit. The many species of mongooses are mostly ground-
living hunters and will tackle almost any form of prey that is
not too large. It is quite true, for example, that the Indian
mongoose is capable of killing a cobra. However, some of the

mongooses are not quite as formidable. An example is the kusimanse of western and central Africa, which feeds on insects, other invertebrates and fruit. The genets and the civets are superficially cat-like, but they have shorter limbs and longer muzzles than cats. They usually inhabit forested areas, and many of them climb well. One member of the family, the otter-civet of southeast Asia, has become adapted for swimming. Because of its lengthy geographical isolation, Madagascar is the home of a number of species which are not found elsewhere. These include several kinds of mongoose, and the largest member of the family, the rare fossa, which is allegedly very fierce.

The family Hyaenidae contains four living species. Superficially these are like dogs, for they are digitigrade and long-legged. However, they are not powerful runners and have weak-looking, sloping backs. Besides this they have fewer molar teeth than the dogs. The spotted and the brown hyenas are found in Africa, and the striped hyena in both North Africa and southern Asia. Hyenas do not always kill their own food, although the largest species, the spotted hyena, is capable of killing a man. They usually feed on carrion and are able to crack even the largest of bones. The fourth species is the aardwolf of southern Africa. Unlike the others this species has weak jaws and feeds almost exclusively on insects.

Spotted hyena;
Africa

The cats are in many ways the most highly evolved of the land carnivores. Including the domestic cat there are about 36 species of them, and they make up the family Felidae. They are digitigrade, and all of them except the cheetah have fully retractile claws. These claws make formidable weapons and also enable all except the heaviest members of the family to climb trees well. The cats also bite powerfully, for they have relatively few cheek teeth and therefore have short jaws. The cats are among the most purely carnivorous members of their order. Essentially they are stealthy hunters, slinking up on their prey and pouncing from as short a distance as possible. Most of them hunt on their own, but some populations of lions live in prides and may combine to hunt as a team. If cats cannot get very close to the prey because, for example, there is no cover, they may sprint over the intervening space, but they never run far, and are not running hunters in the way that the typical dogs are. The caracal lynx and the cheetah, both of Africa and southern Asia, are examples of species with marked sprinting ability, and over a short distance the lightly built and long-limbed cheetah can reach speeds of up to 65 m.p.h.

Wild cats are found in almost all parts of the world except the Australasian region. In any single area there may be a range of species. For example, in southeast Asia there are the tiger, the slightly smaller leopard, the medium-sized

Cheetah;
Africa and India

Walrus;
Arctic seas

clouded leopard, and other still smaller species, including
the leopard cat. Each of these is adapted for a slightly different mode of life. The tiger can deal with the largest of prey,
the leopard is better adapted for climbing and the smaller
species can live on smaller prey which would be of no use
to their larger relations.

The members of the sub-order Pinnipedia are only distantly
related to the rest of the Carnivora. There are three families, and
one of these, the Odobenidae, contains only the walruses.
Whether these constitute one species or two is not certain, but
the walruses of the North Atlantic and those of the North
Pacific do not differ from each other very much, so probably
both populations belong to the same species. Walruses live
in and around cold, shallow seas. They are gregarious and
have been seen in herds numbering several hundreds. Like
all other pinnipedes they have webbed, flattened limbs and
are insulated by means of a thick layer of fat in the skin. The
hind limbs of a walrus can be turned forward under the body
for movement on land, but compared with the seals and sea
lions, walruses are clumsy swimmers. They feed mainly on
mollusks which they dredge up from the sea bed. A large
male walrus weighs well over a ton.

Sea lion;
most oceans

The 12 species which make up the family Otariidae are sometimes collectively known as the eared seals, for unlike the true seals they have small but distinct external earflaps, or pinnae. They have long, webbed forelimbs, and these are their chief means of propulsion when swimming. In the water they can reach speeds of up to 17 m.p.h. On land the hind feet can be tucked under the body so that the eared seals can move with a clumsy galloping motion. The family is widely distributed in the world's major oceans, being found almost everywhere except in the North Atlantic and the warmer parts of the Indian Ocean. They prefer coastal waters.

During the breeding season they congregate on shore in large herds. The males, which are much larger than the females, each attempt to establish a small territory within which they jealously guard their harems. The females each give birth to a single young, and in most species mating follows within a few days. The gestation period is lengthy and occupies the remainder of the year so that all of the business of reproduction, for which dry land or very shallow water is needed, can be accomplished during one short part of each year. The young usually accompanies the mother until the next breeding season and is suckled for most of that time. While at sea, the adults feed on squid and fish. The cheek teeth are pointed to grip the slippery food. Members of this group include the fur seals, which are hunted by man for their skins, and the sea lions, one species of which, the Californian sea lion, is well known to most people as the performing seal.

The true seals form the family Phocidae, which contains about 18 species. They lack earflaps, have short forelimbs and swim mainly by means of their hind limbs. On land the hind feet cannot be moved forward to a position under the body, and the seals have to pull themselves along by means of the forelimbs only. Through evolutionary convergence, they bear a considerable resemblance to the eared seals. They have a wide distribution.

In size they range from the elephant seals, a large male of which may weigh well over three tons, to the Baikal seal which weighs about 170 pounds. The Baikal seal is the only species which is confined purely to fresh water. It is closely related to the Caspian seal and to the ringed seal of the Arctic. Some seals are solitary, but others are gregarious, especially at the breeding season when their behavior is very much like that of the eared seals. Some, like the crab-eater seal of southern oceans, feed on large planktonic animals, but most feed on fish and squid. The leopard seal of the Antarctic feeds on penguins, other sea birds and smaller seals.

Harbor seal;
northern oceans

The aardvark

It is remarkable that the members of ancient and declining orders of mammals often become adapted as anteaters. Examples are the egg-laying echidnas, some of the edentates, the pangolins and the aardvark. Members of each of these groups have separately evolved comparable adaptations. These include powerful claws with which to open ant hills, slender skulls with jaws which bear few, if any, teeth and very long tongues. The aardvark is in all of these respects a typical anteater, and it is the sole species in the family Orycteropodidae which, in turn, is the only family in the order Tubulidentata. It was once thought to be related to the South American edentate anteaters, but it now seems more likely that it is descended from some of the early ungulates.

The aardvark inhabits open country, bush and occasionally forests in Africa south of the Sahara. Its name is Afrikaans and means 'earth pig'. It is indeed about the size of a pig, measuring about seven feet from head to tail and weighing up to 140 pounds. It is digitigrade and has four digits on the forelimbs and five on the hind limbs. Despite its relatively long limbs it cannot move at any great speed, and when it is above ground its main defenses are its tough skin and its ability to rear on its hind legs and lash out with its claws. These tactics are often successful against such predators as lions, hunting dogs and large pythons. However, its most effective defense is provided by its remarkable burrowing ability. Digging with its forefeet and removing the excavated soil with the hind feet and the tail, it can rapidly construct a burrow even in the hardest of ground. The aardvark is predominantly nocturnal, and its burrow provides it with shelter during daylight, although it sometimes suns itself at the entrance. Although aardvarks tend to be solitary, sometimes the burrows of a large number of individuals are found in quite a small area. At night they feed on termites, ants, locusts and possibly some plant material.

The gestation period of the aardvark is not known. The female bears one or very occasionally two young, and in the first weeks of their lives they remain within the safety of the burrow. At about six months the young are sufficiently developed to start making their own burrows.

Aardvarks;
Africa; about
7 ft.

The elephants

Although its members were more numerous in the past, the order Proboscida now contains only one family, the Elephantidae. Most zoologists would probably agree that there are only two species, the African elephant of Africa south of the Sahara, and the Indian elephant with a range which extends from India to Sumatra and Borneo, but some are inclined to regard the elephants of Africa as members of two different species. However, the small elephants to be found in some African forests which make up the second of these are probably merely immature members of one race of African elephants and not really a distinct species.

It is well known that African elephants have larger ears than Indian elephants, but there are a number of other external differences. The African elephant tends to be a taller, lankier animal, while the Indian has a more rounded outline. African elephants are also usually heavier. A large bull may be 11 feet tall and weigh well over six tons. All adult African elephants usually have tusks, those of the bulls being the larger, but in the Indian species only the bulls have tusks. The cows have only small tusks which scarcely protrude beyond the lips, and they are usually hidden by the trunk.

African elephant

The African elephant has two finger-like processes at the end of the trunk, and the Indian elephant has only one. The African elephant normally has four nails on its front feet, but the Indian has five; both species have either three or four nails on the hind feet.

As they are the largest of land mammals and feed upon plant matter which is difficult to digest and low in food value, elephants have enormous appetites. They may spend up to 18 hours a day in feeding, resting only in the middle of the day and for part of the night. The trunk consists of the nose and the upper lip and is highly adapted primarily as a food-gathering organ. With considerable dexterity it conveys branches, leaves, fruits and shoots to the elephant's mouth. Additionally, elephants use their trunks for drinking, filling the nostrils with water before squirting it into the mouth, for bathing and dust-bathing, as a snorkel tube when swimming and as a weapon. The cheek teeth are adapted for grinding large amounts of tough food, while despite their shape the tusks are really incisor teeth.

The gestation period is variable, but it is between 18 months and two years. There is usually a single young. Like all other mammals young elephants suckle with their mouths. They grow up at about the same speed as human beings, and like man they live for about 70 years.

Indian elephant

137

Tree hyrax;
Africa

The hyraxes

The order Hyracoidea contains one family, the Procaviidae, with perhaps six species. These are the hyraxes. Superficially they look like large rodents, but in fact the closest relations which the hyraxes have are probably the elephants. However, even this relationship is a remote one. By means of their fossils the hyraxes' history as a separate group can be traced back for many millions of years. It seems that they first evolved in Africa, and they have never spread very far. Even today only one species occurs outside Africa, being found in southwestern Asia, including Israel and parts of Arabia.

In size, the hyraxes are about as large as rabbits, but their short tails are inconspicuous, and they have only short ears. Although the limbs are plantigrade they bear hoof-like claws. There are four digits on each forelimb and three on each hind limb. The bare pads on the soles of the feet are richly supplied with sweat glands which keep the feet slightly damp and enable them to keep a grip even on smooth, sloping surfaces. This is useful, for although some species live on the ground, they are found of scrambling among steep rocks while others climb trees. The scent produced by these glands is probably also important as a social signal. The position of the glands ensures that the hyraxes, like some of the carnivores and ungulates, leave a trail of scent wherever they go. Ringed with long guard hairs among the fur on the middle of the back of the body is another scent gland. Obviously smell plays an important part in the social like of the hyrax, but its hearing and eyesight are also keen.

The diet is purely herbivorous and consists of grass, leaves and shoots. The upper incisors of the adults are long and the cheek teeth are adapted for grinding plant material, although not so well as those of some of the rodents or ungulates. Many herbivorous mammals have a large sac, the caecum, leading from the alimentary canal. This structure assists in the digestion of cellulose, an important plant material. The hyraxes are unique in having not one caecum, but three.

Tree hyraxes are nocturnal and lead solitary lives, but the gregarious ground-dwelling species are usually most active in the early morning and the evening. For such small animals they have the relatively lengthy gestation period of eight months. The litter consists of from one to three young, and they are born with their eyes open.

Rock hyrax;
Africa

The sirenians

The members of the order Sirenia, or sea cows, are, after the cetaceans, the most highly adapted of all mammals for an aquatic life. It is true the seals and sea lions are more rapid swimmers, but the sea cows are more completely aquatic, for they even give birth to their young in the water. In some ways they resemble the whales for they have horizontally flattened tail flukes, their skins are almost devoid of hair, the five digits of their forelimbs are completely webbed to form flippers, the hind limbs are absent and only a few vestiges of the pelvic girdle remain. However, the sea cows are not at all closely related to the whales and are the descendants of an ancient group of herbivorous mammals. They may have some distant connection with the elephants. The surviving sea cows average about eight feet in length and

Dugong; Ethiopian
and Oriental waters

weigh about 500 pounds. They inhabit coastal waters and rivers in warm climates and feed on seaweed and other aquatic plants. There are two families.

The family Dugongidae contains only a single living species, the dugong. This has a range which extends from the Red Sea to the coasts of Madagascar and northern Australia. The dugong had a deep notch in the middle of the hindmost edge of its tail fluke, and males have tusk-like incisor teeth. However, the lower incisor and some of the cheek teeth are very frequently absent. Almost certainly it was the dugong which gave rise to sailors' tales of mermaids.

Manatees; waters of southeastern
United States to South America

The notched tail suggests that of a fish, while the female
dugong's habit of clasping her single young to her breast
is undoubtedly mammalian. Any suggestions of beauty
according to conventional human standards can only have
been produced by imagination heightened after many days
at sea. A second member of this family was discovered in
the North Pacific during the eighteenth century, but was
hunted to extinction soon afterward. This was Steller's
sea cow, a large species up to 30 feet long.

The family Trichechidae contains the manatees, of which
there are three species, one from the coasts and rivers of
West Africa and two from those of eastern North America
and South America, and the Caribbean. One of these American species is found only in fresh water in the Amazon and
Orinoco River basins. Manatees have tails which lack a notch
and have no incisor teeth when adult. They have a large and
very variable number of cheek teeth. They spend much of
their time feeding, using their muscular mobile lips to crop
the vegetation and surfacing to breathe about once every
ten minutes. They are usually solitary. All other living mammals, except some of the sloths, have seven cervical (neck)
vertebrae, but the manatees have only six.

Malay tapir with young; Oriental

The odd-toed ungulates

The ungulates or hoofed mamals include the most successful of large herbivorous mammals. There are two orders of which the Perissodactyla (odd-toed ungulates) was the first to become successful. Its members differ in a number of details from the members of the order Artiodactyla (even-toed ungulates), but externally they can best be distinguished by their limbs. Like the artiodactyls they have claws which have become modified to form hoofs so that the limbs are as long as possible to facilitate running, and the number of digits has become reduced to make the feet as light as possible. However, in the perissodactyls the axis of each foot passes through the third digit. The digits toward either side of each limb accordingly become unimportant and tend to be lost so that there are almost always three digits, or one digit on each limb. For this reason, the perissodactyls are often called the odd-toed ungulates. More importantly the perissodactyls have simple stomachs and rely upon a large caecum for the digestion of cellulose. Their alimentary canals are not as efficient as those of the artiodactyls, and probably for this reason they have been slowly declining in numbers for the past thirty million years. Only 15 wild species, divided into three families, now remain.

The family Tapiridae contains the most primitive living members of the order. There are four species of tapirs, three of which inhabit South America while the other is found in the Oriental region from Burma to Sumatra. Tapirs are forest-loving animals and because of this have relatively short limbs. On their forelimbs the thumbs are absent but the little fingers remain, so that there are four digits. There are only three digits on each hind limb. They are heavily built, and are about three feet high at the shoulder. They are nocturnal and solitary. South American tapirs are gray-brown, but the Malay tapir has bold black and white markings. All young tapirs are striped. Tapirs keep to thick cover and never go far from water, in which they swim with ease. They have very mobile upper lips and noses which serve to tuck branches into the mouth while browsing.

Five species of the family Rhinocerotidae survive, two in Africa and three in the Oriental region. Rhinoceros horns are unique in that they grow from the skin, consist only of keratin and grow on the mid-line of the body. Both African species and the little Sumatran rhino have two, while the Indian and Javan rhinos have only one horn. In addition, the Asian rhinos have razor-sharp incisor teeth they can use as weapons. Rhinos are solitary and tend to be peaceful creatures, except for the African black rhino which has an unaccountably bad temper. All rhinos are herbivorous. Most of them are browsers, except for the white rhino, which is adapted for grazing.

Black rhinoceros;
Ethiopian

143

Mongolian wild horse, or Przewalski's horse; central Asia

The family Equidae contains the horses, asses and zebras of which there are six surviving wild species and one domesticated one. The fossil history of this family is unusually clear. As is well known, the ancestral horses were no larger than dogs and had several digits on each limb. As the group became better adapted for swift running over firm plains, its members became larger and their lateral digits diminished until there was only a single digit and a single hoof formed from the claw of the middle finger or toe upon each leg. Some of the ancestral horses inhabited North America, but the group now occurs naturally only in the Old World. Until man intervened, wild horses inhabited the plains of central Europe and Asia; wild asses came from warmer areas further to the south in both Asia and Africa; and zebra came from still further south in Africa. The domestic horse was first tamed in Asia some thousands of years ago and is regarded as belonging to a separate species from the very few wild horses which remain in Mongolia. The domestic ass is thought to belong to the same species as the African wild ass.

The Mongolian wild horse is stockily built and has a short, stiff mane. It was first discovered in the nineteenth century by a Polish traveler named Przewalski. In appearance it is very like the extinct European wild horses which we know from cave paintings found in France and Spain.

144

The two species of wild ass, one from Asia and one from Africa, are more lightly built with very slender limbs. Although there is only one species of wild ass in Asia, a number of different geographical races with different common names are recognized. For example, the kiang is the largest form, while the onager is smaller and more slender. Three species of zebra survive. The largest of these is Grevy's zebra, which inhabits hot, dry areas of the 'horn' of Africa and is marked with numerous thin stripes. The common zebra is found in East Africa from the Sudan southward. It is distinguished by varying patterns of stripes. The smallest zebra is the mountain zebra of southern Africa which has a characteristic 'gridiron' pattern of stripes on the rump.

All the surviving members of the horse family are quite closely related. Domestic horses have been successfully crossbred both with donkeys and zebras. All members of the family are gregarious, sometimes living in quite large herds. They are active both by day and by night. Grass is their chief food, and in cropping it they use both their lips and their sharp incisors. The canine teeth are small or absent, and the cheek teeth form a continuous row. A single young is born after about 11 months gestation. The life span may exceed 20 years.

Common zebra;
Ethiopian

Onager;
Asia

The even-toed ungulates

The members of the order Artiodactyla are sometimes called even-toed ungulates. This is because they have either four or two hoofed toes on each foot. In either case the axis of each foot passes between the third and the fourth digits of each limb, making these two digits the largest and strongest. The second and fifth digits are usually also present, although in most cases they are much reduced in size, but the thumb and the big toe are always absent. Like the perissodactyls, they are typically large herbivores. The most highly evolved artiodactyls have complex chambered stomachs and are able to chew the cud. As a result of these adaptations they are able to make more efficient use of plant food and are therefore a successful group. In the wild, they are found in all parts of the world except for some remote islands and the Australasian region.

The family Suidae contains the pigs, all of which are stoutly built and are short-legged. They are a primitive group, and have four well-developed toes on each limb. They cannot chew the cud. Wild pigs first evolved in the Palaearctic region, and they have never colonized the New World. Most of them live in forests or bush country and are omnivorous. The canines are curving and tusk-like and

Wild boar;
Palearctic

Babirusa; southeast Asia

are at their largest in the babirusa of Celebes. Other species include the wild boar, which inhabits much of the Palaearctic region and part of Africa and which is the ancestor of the domestic pig. The bush pig, forest hog and wart hog inhabit Africa. There are eight species.

The two species contained within the family Tayassuidae, the collared peccary and the white-lipped peccary, are superficially like the pigs and are adapted for much the same mode of life, but the two families are not very closely related. Peccaries are found only in the New World from the southwestern United States to Argentina.

Collared peccary; southwestern United States to Argentina

The family Hippopotamidae has only two living species and they differ greatly in size. The hippopotamus, which is widespread in Africa, may measure up to 13 feet long and weigh over three tons, while the rarer pygmy hippopotamus of the coastal forests of West Africa is about as big as a large pig. Both are purely herbivorous and have large, peg-like incisor teeth and cheek teeth with rounded cusps. Although they cannot chew the cud, they have complex stomachs which are divided into three chambers.

The pygmy hippo is a shy animal and usually remains among thick cover. It lives singly or in pairs and is usually most active at night. Although it is fond of water, it spends most of its time on land. The larger species is truly amphibious, for it spends much of the day in shallow rivers and pools, dozing and feeding on water plants. As is often the case with amphibious vertebrates, it has the nostrils, eyes and ears so arranged that they are just above water when all the rest of the body is submerged. It often goes right under the water, walking on the river bed with its nostrils closed; it can remain below water for as much as ten minutes. At night it grazes on land. It is gregarious and usually lives in herds of a dozen or more. Both species of hippo normally give birth to a single young.

The family Camelidae first evolved in North America but is now extinct there, and wild members of the family are now found only in Asia and South America. Although domesticated camels have run wild in several parts of the world, the only truly wild humped camals are the Bactrian camels of the Gobi Desert. These have two humps, and belong to the same species as domesticated two-humped camels. The one-humped Arabian camel, or dromedary, is known only as a domesticated animal. In South America there are four more humpless species belonging to the family. These are the guanaco and the vicuna, both of which are wild inhabitants of the high plateaus of the Andes, and the domesticated llama and the alpaca. They are purely herbivorous and have soft, mobile lips and three-chambered stomachs.

Hippopotamus (*left*);
Africa

Vicuna (*right*);
Neotropical

Bactrian camels (*below*);
central Asia

149

The remaining artiodactyl families are all able to chew the cud. The four species of the family Tragulidae are probably more closely related to the deer than to any other family. They are commonly known as chevrotains and are small animals, about a foot high at the shoulder. They are shy, solitary and nocturnal, and they usually inhabit dense forests not far from water. One species, the water chevrotain, has a range which includes West Africa and the Congo Basin, and the other three occur in Asia from India to Borneo. They have no horns or antlers but their upper canines are long and sharp, especially in the males.

About 40 species of deer make up the family Cervidae. In most species the males bear antlers, but in the primitive musk deer of central Asia and in the little water deer of China males have long upper canines instead.

Antlers differ from horns in that they consist of bone without an external covering of keratin; they are branched, and deciduous. Typical artiodactyl horns are permanent structures, but male deer use their antlers in the fighting which accompanies the mating season and lose them afterward. Soon afterward a pair of antlers starts to grow from the bare

Water chevrotain;
West Africa and Congo

Musk deer; Asia

150

pedicels on top of the head. While they are growing, the antlers are covered with furry skin beneath which there are blood vessels. At this time stags are not at all pugnacious, for it would be very painful for them to use their antlers as weapons while they are skin-covered. When the antlers are full-grown the skin dies and falls off, and only bare bone is left. Each year until it reaches full maturity—at about ten years of age in the case of the larger species—a male grows a slightly larger and more branched set of antlers, but it is not true that there is one branch for each year of the stag's age. A really fine set of antlers on a large deer such as a wapiti might weigh as much as 40 pounds. Female deer do not fight each other during mating season and therefore have no need of antlers. The only species in which both males and females bear antlers is the caribou, and the reason for the females being thus adorned in this single case is not at all clear.

Some small deer are solitary, but most live in herds. All of them are herbivorous and feed on grass, bark or leaves and, in the case of the caribou, lichens.

Red deer; Palearctic

Giraffe;
Ethiopian

The family Giraffidae has never occurred outside the Old World and is now confined to Africa south of the Sahara. There are only two species. These are the giraffe, the tallest living mammal, which has been known to attain a height of over 18 feet, and the okapi. Both are long-necked, but like almost all other mammals they have only seven cervical vertebrae. They have short horns which are covered with hairy skin rather than with keratin. They have only two hoofed digits on each foot. Both species are browsing herbivores with long tongues. The giraffe can stick its tongue out for well over a foot.

Giraffes live in herds of up to 20 individuals. Each herd normally contains one mature male, a number of females and their young. They live in open, sparsely treed grassland and can run as fast as a racehorse. They breed at all times of the year. The gestation period is about 15 months, and there is almost invariably a single young, although twins have been recorded. The female gives birth while in a standing position, so the young giraffe begins its indepen-

Okapis;
central Africa

dent career by falling from a considerable height. Despite this it can rise to its feet and suckle within minutes of being born. It is about six feet tall and can run before it is 24-hours old.

The okapi lives in the forests of the Congo and is nocturnal, shy and solitary. It is about five feet tall at the shoulder.

The prongbuck is the sole surviving member of the family Antilocapridae. It is sometimes called the American antelope, but it is not a true antelope. Both males and females have short, simply branched horns which have a conventional core of bone growing from the skull, and a unique outer sheath of horn which is lost and grown again each year. It has the same dental formula as the giraffes, and like them it has only two toes on each foot, but it is not very closely related to them. Prongbucks live in dry areas of western North America, and their range has been greatly reduced by human activities. During the summer they live alone or in small herds, but larger herds form during the winter. They feed on grass, sagebrush and succulent plants.

Prongbucks;
western North America

Sable antelope;
Ethiopian

The most successful family of large mammals is the Bovidae. It contains over 100 species and occurs naturally in all parts of the world, except the Neotropical and Australasian regions. Domesticated members of the family, such as cattle and sheep, have been taken to virtually all inhabitable countries. Members of the group all have four toes on each foot, although the lateral ones are small. True horns are often present. These horns are permanent with a core of bone which grows from the skull and are covered by a layer of keratin. They are herbivores which chew the cud, have four-chambered stomachs and have a very efficient digestive system. The cheek teeth have ridged surfaces ideal for grinding plant material; they are designed so that they are rendered more efficient by wear.

Included in the family are a large number of species of

Rocky Mountain goats,
northwestern North
America

Water buffalo;
Oriental

antelopes. The smallest of these is the royal antelope of West African forests which is only ten inches tall. Slightly larger, more graceful forms are the gazelles of the dry plains of Africa and Asia. The largest of the antelopes is the giant eland of West Africa and the Sudan which may be over six feet in height at the shoulder. In some species of antelopes, both the males and the females bear horns.

The heavyweight members of the family include the European and North American bison. The name 'buffalo' is best reserved for equally heavy species from warmer climates which have thinner coats and flattened horns. Examples are the water buffalo of southern Asia, and the cape buffalo of Africa south of the Sahara. Other species of cattle include the banteng of the Oriental region, the yak of central Asia, the zebu and domestic cattle.

The wild sheep and goats from still another group. They are adapted to exploit the fact that a hoofed foot is small and can gain a foothold even in small crevices. The wild goats in particular often inhabit very wild and rocky country in Asia and North America. The domestic goat was derived from a wild species of Asia, while the domestic sheep probably had ancestors very like the wild mouflon which inhabit Sardinia and Corsica today.

BOOKS TO READ

For general introductions to the subject, the following titles are recommended and are usually available from bookshops and public libraries.

Mammals of the World (2 vols.). Ernest P. Walker, Johns Hopkins Press, 1964.

Recent Mammals of the World. Sydney Anderson and J. Knox Jones, Jr. (eds.) Ronald Press, 1967.

The Mammals. Desmond Morris. London: Hodder & Stoughton, 1964.

The Natural History of Mammals. François Bourlière. Knopf, 1954.

Living Mammals of the World. Ivan T. Sanderson. Doubleday, 1961.

The Monkey Kingdom. Ivan T. Sanderson. Chilton, 1963.

The Mammals of North America (2 vols.). E. Raymond Hall and Keith R. Kelson. Ronald Press, 1959.

A Handbook of Living Primates. J. R. Napier and P. H. Napier. Academic Press, 1967.

Whales. E. J. Slijper. London: Hutchinson, 1962.

Bats. Glover M. Allen. Dover, 1962.

A Field Guide to the Mammals. William H. Burt and Richard P. Grossenheider. Peterson Field Guide, Houghton Mifflin, 1964.

A Field Guide to the Mammals of Britain and Europe. F. H. Van Den Brink. Peterson Field Guide, Houghton Mifflin, 1968.

Mammals of North America. Victor H. Cahalane. Macmillan, 1947.

Biology of Mammals. Richard G. Van Gelder. Charles Scribners' Sons, 1969.

A History of Land Mammals in the Western Hemisphere. William B. Scott. Hafner, 1962.

The Mammals. Time-Life Books, 1969.

PLACES TO VISIT

American Museum of Natural History, New York City.
Smithsonian Institution, Washington, D.C.
Field Museum of Natural History, Chicago, Illinois.
Los Angeles County Museum, Los Angeles, California.
Bronx Zoo, New York City.
Staten Island Zoo, New York City.
Philadelphia Zoo, Philadelphia, Pennsylvania.
Franklin Park Zoo, Boston, Massachusetts.
National Zoological Park, Washington, D.C.
Lincoln Park Zoo, Chicago, Illinois.
Milwaukee County Zoo, Milwaukee, Wisconsin.
San Diego Zoo, San Diego, California.
San Francisco Zoo, San Francisco, California.

INDEX

157

OTHER TITLES IN THE SERIES

The GROSSET ALL-COLOR GUIDES provide a library of authoritative information for readers of all ages. Each comprehensive text with its specially designed illustrations yields a unique insight into a particular area of man's interests and culture.

NOW AVAILABLE

PREHISTORIC ANIMALS
BIRD BEHAVIOR
WILD CATS
FOSSIL MAN
PORCELAIN
MILITARY UNIFORMS 1686–1918
BIRDS OF PREY
FLOWER ARRANGING
MICROSCOPES & MICROSCOPIC LIFE
THE PLANT KINGDOM
ROCKETS & MISSILES
FLAGS OF THE WORLD
ATOMIC ENERGY
WEATHER & WEATHER FORECASTING
TRAINS
SAILING SHIPS & SAILING CRAFT
ELECTRONICS
MYTHS & LEGENDS OF ANCIENT GREECE
CATS, HISTORY—CARE—BREEDS
DISCOVERY OF AFRICA
HORSES & PONIES
FISHES OF THE WORLD
ASTRONOMY
SNAKES OF THE WORLD
DOGS, SELECTION—CARE—TRAINING
MAMMALS OF THE WORLD
VICTORIAN FURNITURE AND FURNISHINGS
MYTHS & LEGENDS OF ANCIENT EGYPT
COMPUTERS AT WORK
GUNS

SOON TO BE PUBLISHED